GOD

IS IN THE

DOUBT

TOUCH

handcrafted services

Advance Praise

"*God Is In the Doubt* is an honest and gut-wrenchingly real observation of grief and loss in starting and building a family. People who have struggled to build their family, and have tried everything, might even find comfort in knowing that their story isn't entirely unique and the pain is felt by too many too often. David Lloyd's honest and transparency shines through in a way that is totally familiar. It rips open wounds and leaves them bare while pointing back to a foundational faith that can't be ignored, overlooked, or even unexamined."

Greg Sykes, Creative and Digital Director at Sykes Marketing & Design

"David Lloyd's book, *God Is In the Doubt,* is a raw and unapologetic story of anger and doubt and faith. To call it his testimony or a confession would be trite. It is David's life. If you can read it without feeling anything, then you're not paying attention. David shares every tragedy and all the attendant emotions without glossing them over or assuaging them with cheap platitudes. He puts all of it on the page as it was and he doesn't ask you to agree with him. This is his revelation, how he found out that God wasn't afraid of his doubt or his fear or his anger. This is David telling you how he learned that God was always with him, especially when he was shaking his fist at Him. If you come to this book and you already have faith, David will challenge it. If you come to it and you do not have faith, but you do have an open mind, David might help you find it. Either way, as David shows us, it's okay to have questions."

Jamey Potter, Author of In the Presence of Strangers

"We live in a culture that is more than ready to throw stones and so easily accept the advice of Job's wife. But instead of cursing God after what seems like a barrage of worst of times, the Lloyds strengthened their faith and are an inspiration to me and countless others. I remember getting the phone call. There are a million and one reasons why I might not immediately answer my phone. On that day, in that moment, there was nothing more important and I will forever regret first hearing about Caleb from David's voicemail. Thankfully, the grace and compassion of the elders, leaders, and congregants of our small church, as well as my wife who didn't take the trip with us, poured over David and Donna in one of their darkest hours. David has heard me preach that faith is like love - it is either real or it isn't. There is no losing love or faith, only not having the genuine version of it to begin with. Watching and walking with the Lloyds as they grieve has reminded me what real faith looks like. The greatest tragedies will never waver the most basic genuine faith. Sure, we may have doubts and questions, as even I, a full-time pastor, still often do. But having friends like David and Donna encourage and influence me to strengthen my grip on my faith even more. I love the Lloyds and believe their story will bless you as it has me in real life and in again in reading these pages."

Will Goodwin, Worship Gatherings Pastor at Crosspoint City Church

"David shares his faith journey with authenticity and vulnerability, rarely seen in the world today. From the grief and heartache of multiple miscarriages, to the tragedy of suicide, and multiple serious accidents, he had every reason to give up on God. Despite his many trials and tribulations, David intensely pursued the reason for the hope we all should have within us as Christians. Through his scientific research, burdensome life circumstances, and through much prayer, David gained an unwavering faith in God. His story will certainly touch many lives."

Jay Torrence, Founder of PersistSEO

"A vulnerable story of loss and redemption. David has written from the heart, and it kept me turning the pages."

Faith Griffin Sims, Author

"David uses his background in biblical studies and mathematics to walk us through a life that contains an unbearable amount of loss and pain. In a world where most people would usually give up on their faith, David uses mathematics to hold firm to his faith, knowing that his God is there.

The words you will read in this book are sometimes raw, can be difficult, and always genuine. You get to peek into David's life and see what it means to hold on to your faith, even in the midst of an external force that never seems to relent from its sending negative situations.

If you are actively processing deep loss, looking for a way to hold on to your faith, and/or need more ways to bring hope into your life, read this book."

David Webber, Entrepreneur

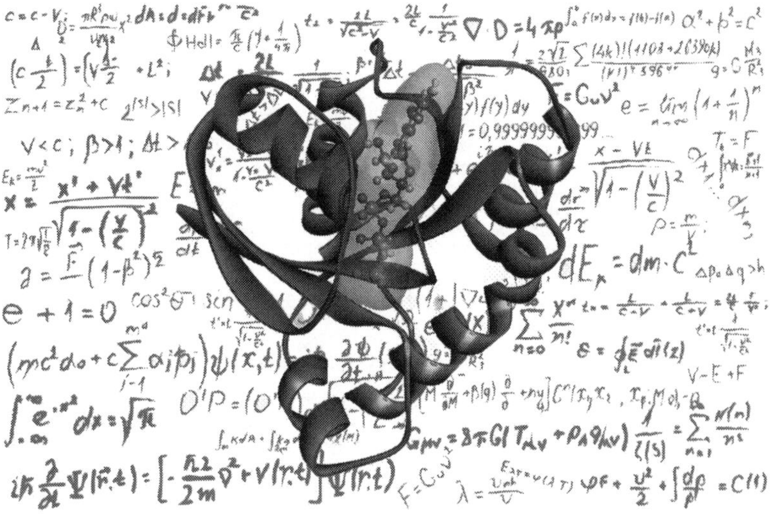

GOD

IS IN THE

DOUBT

memoir on keeping faith through loss
David A. Lloyd

TOUCH

handcrafted services

This is a work of creative nonfiction; a story based on my life. As all stories go, it may be inaccurate. I have tried to recreate events, locales and conversations from my memories, but, as we all know, memories are deeply flawed. Some names and identifying details have been changed to protect the privacy of the people involved. Any conversations are from my recollections and convey the idea, but not necessarily the word-for-word transcript.

God Is In the Doubt

Scriptures marked (KJV), or otherwise not noted, are taken from the KING JAMES VERSION, public domain.
Scripture quotations marked (YLT) are taken from the 1898 YOUNG'S LITERAL TRANSLATION OF THE HOLY BIBLE by J.N. Young, public domain.

Requests for information should be sent to the publisher:
TOUCH: Handcrafted Services, Inc.
151 E. Marietta St, Suite 3, Canton GA 30114
info@thcs.co   www.thcs.co

ISBN: 978-1-7371471-0-7 (print)
ISBN: 978-1-7371471-1-4 (ePub)
LCCN: 2021908978

To Alex, Caleb, Madilyn, and Emily

"If you would be a real seeker after truth, it is necessary that at least once in your life you doubt, as far as possible, all things."
— René Descartes

Contents

Foreword

Having lost his sweet Helen Joy, C. S. Lewis once said, "Not that I am (I think) in much danger of ceasing to believe in God. The real danger is of coming to believe such dreadful things about Him. The conclusion I dread is not 'So there's no God after all', but 'So this is what God's really like. Deceive yourself no longer.'" Those are such honest words. Most people have such a shallow understanding of God that they fear speaking so transparently about their true feelings and thoughts, having endured great loss while clinging to a shrinking faith.

David Lloyd has put his emotions, thoughts, and faith in a fishbowl, which is this very book. *God Is In the Doubt* is a reading experience like unto a visit to an Amusement Park. You'll reach your hands toward the heavens as you take the drop with David on the rollercoaster of the human experience. The emotional G's will make you lose your breath but you'll catch a grip on hope as David articulates how God breathed on him and his family during a time of great loss. Fear will grip your heart as you experience what felt to me like a haunted house; grief being the ghosts that taunted and tormented David. As you go through each chapter, you'll read of David's ride on the Not So Merry-Go-Round of doubt and confusion, twins that wage war on our faith. But like any ride, there is the moment you get on and the moment you get off of it. So it is with each chapter and the experiences you'll have. David doesn't leave the reader on a ride spinning out of control, rather he helps to secure a two fisted grip on the sovereignty of a God who can handle our doubts.

While it is true we love to view a sunrise, the stars infuse our hearts with tranquility. Yet, the only way to view the heavens is in the dark. David and his sweet family have endured a dark storm few are able to weather. Through the fallout of the human experience, they were cast into a furnace of affliction. In that churning abyss, they walked in such a way that when you stop to peer into their furnace by reading this book, you will clearly see the fourth man walking with them. The Bible narrative is very clear that faith makes things possible, but it doesn't make it easy. God said everything would work out not that everything would make sense. The loss of a loved one can put us all in a state of painful confusion. The most painful goodbyes are the ones that are never said and never explained. Jeffrey Holland said, "Faith draws the poison from every grief, takes the sting from every loss, and quenches the fire of every pain." Did you see it in Holland's quote? It is something we know to be true, but when pain comes, we hope it was a lie. The overlooked and underlining key to overcoming pain is by enduring pain. David breaks the code to help us navigate the things that break us.

Holly Kohler described the healing process following the suicide of a loved one. She said, "A person never truly gets 'over' a suicide loss. You get through it. Day by day. Sometimes it's moment by moment." This book is a powerful tool to help you keep pressing on in the moment by moment recovery. Though God saves us sinners from the pit of Hell, He does not save us from living Hells. Jesus

Christ said, "In this world you will have tribulation." We will suffer and shed tears. Viktor Frankl consoles our hearts by saying, "But there was no need to be ashamed of tears, for tears bore witness that a man had the greatest courage, the courage to suffer." David and his family have suffered greatly, and in so doing, wrestled with understanding God, his purposes, and plans. You'll appreciate the candor and straightforward talk David gives about faith in the midst of loss. Elie Wiesel once said, "I have not lost faith in God. I have moments of anger and protest. Sometimes I've been closer to Him for that reason." David's insights help the reader experience the same sort of closeness to God when your circumstances seem to say he's so very far away.

Rob Liano writes, "The sorrow we feel when we lose a loved one is the price we pay to have had them in our lives." David and his family have and are still paying a great price. How can we pay when we don't have enough in our emotional bank account to cover the cost? Elizabeth Kubler-Ross says, "The reality is you will grieve forever. You will not 'get over' the loss of a loved one; you will learn to live with it. You will heal and you will rebuild yourself around the loss you have suffered. You will be whole again but you will never be the same again. Nor should you be the same nor should you want to." The passage you'll make through this Book will empower you to embrace the fact that you'll never be the same again, but that doesn't mean you won't live again.

Managing grief while exercising faith is a beast as Vicki Harrison puts it, "Grief is like the ocean; it comes in waves, ebbing and flowing. Sometimes the water is calm, and sometimes it is overwhelming. All we can do is learn to swim." *God Is In the Doubt* is a literary swimming school. Some who read may already have great skills, but David's insights will help them refine their abilities to such a degree that they'll be swimming the waters of loss like Gold Medalist Michael Phelps. For others who struggle greatly, the keys and advice in this read will help them move from treading water to a powerful breaststroke in the pool of pain. There will be many who hurt so deeply they've been drowning, and David's Lloyd's wisdom will be a lifeline to help you get back safely to Christ's ship of faith. It's His ship of salvation that made the greatest difference for David and his family. Randy Alcorn says, "If we don't know Jesus, we will fear death and its sting- and we should." Lloyd's contagious, battle tested faith is one that will help all readers to never fear death nor cringe its sting.

I opened my foreword, with a quote from one of mine and David's favorite Authors, and I'd like to close with another one. C. S. Lewis said of grief, "For in grief nothing 'stays put.' One keeps on emerging from a phase, but it always recurs. Round and round. Everything repeats. Am I going in circles, or dare I hope I am on a spiral? But if a spiral, am I going up or down it? How often -- will it be for always? -- how often will the vast emptiness astonish me like a complete novelty and make me say, 'I never realized my loss till this moment'? The same leg is cut off time after time." Anyone missing legs will tell you they have a hard time swimming. The transparency and wisdom of *God Is In the Doubt*, are emotional and spiritual prosthetics with fins for those who have been tossed in the deep end of loss and grief.

Fellow swimmer,
Tony Nolan

~ ~ ~

Tony Nolan is no stranger to the rollercoaster of the human experience. He was born in a mental institution, abused by foster parents and his adoptive parents, raised in poverty in Jacksonville's Sin City, and survived a suicide attempt before placing his faith in Jesus. Tony is the author of the books <u>GASP!</u>, <u>Hurt Healer</u>, and <u>Faith Fuel</u>. He has toured with Casting Crowns and Winter Jam. Tony has comfortably spoken to a few and to 150,000 people in tiny churches and in most of the major arenas in the United States. He is a husband, father, and teacher, but I am pleased to call him a friend. – David A. Lloyd

Preface

When you read or think about events, do you have a mental image?

Maybe I should be more clear. When people would say, "Imagine going up a mountain and seeing the landscape for one hundred miles ...", I would think I knew the general idea of a mountain. I could think about when I was looking out and know the land stretched far to the horizon. I thought this was a mental image. What do you think when someone teaches you a tangent to a circle is a line touching a circle at only one point, like a board balancing on a ball? I know what these objects are. I know what they look like. When taught this math concept I understood the spatial relationship.

A few years ago I discovered most people do see some kind of picture and many experience the vision as if it were real. This shocked me. I do not see an image like looking at a picture or replaying a movie. My imaginings are more spatial relationships, concepts, verbal communications, and sometimes smells. I can not see familiar faces, colors, or even shapes. In the tangent example, I see none of it until I draw it, which I can do. I have a condition called Aphantasia which prevents me from seeing pictures in my mind.

Do not be too surprised while reading this memoir if some memories seem bland or one dimensional. Take some comfort in realizing this is how I see my world. I do not think the world is bland but you may think I do, until the realization hits how I see concepts in spatial terms. My memories are different from what you might expect. I recall concepts, feelings, and ideas better than what was exactly said or exact places. The memories expressed in this book are real as I remember them. Memories are fickle sometimes and we all see events from different views. When I quote someone, the words may have been different but the quote reflects the words and spirit of what I recall.

I quote several scientists and philosophers in this book. Nothing about these quotations imply these people approve of this book or ideas within, nor does it imply I approve of everything these people may promote. This is a story of my journey of personal faith and nothing more. I encourage readers to find these people and read their works. These are people smarter than I.

1

Doorbell

June 2018

A sunny June morning, I took my wife and ten-year-old daughter to church like most Sundays, then came home and changed clothes in preparation to buy a Volkswagen Super Beetle. My love for these cars has not subsided since my first one, a baby blue 1972, when I was in college twenty-nine years earlier. I put the cash in my pocket as the doorbell rings.

Two cops stand on the porch.

Their words send me reeling backward, a gut punch. My back slams into the wall. The two words felt like bricks pummeling me. He's dead.

A piercing wail forces me to turn. My wife has collapsed in the doorway. She tries to rise while also holding our daughter back, who screams. "What's wrong?"

I look at the two cops who just confirmed we are Caleb's parents.

"I'm sorry for your loss."

I do not know what I may appear to the world, but to myself I seem to have been only like a boy playing on the seashore, and diverting myself in now and then finding a smoother pebble or a prettier shell than ordinary, whilst the great ocean of truth lay all undiscovered before me. - Isaac Newton

Lanky with bright white hair, I looked more like a mop than a boy. Always awkward and stumbling often, I kept a knot on my forehead until the age of ten.

Through the teen years, it was common to have bruised arms and legs. My great-grandfather owned and operated a small fishing business. He always encouraged me to catch myself when falling by saying, "Throw out your nets, son!"

For most of my childhood I lived 1,000 miles away from the dock where the salt air lifts the soul. Working on the dock was fun, but I am more of a nerd. Who knows if things would have been different had we not moved. Perhaps reading would not have become a refuge.

The projects were rough for an eight-year-old new kid on the block. Voracious reading allowed an escape from the kids who harassed me for years. The library sat next to the school and I would stop in on my way home from school, sit on the library floor between stacks, and read most anything on the shelves. It was all which home was not—quiet, cool, and peaceful. My reading varied from Hardy Boys mysteries, how-to manuals on wiring homes, physics, science, and history books—or just randomly picked pages in the encyclopedia. I learned how the universe works, Einstein's theories on relativity and work on the atom bomb, anatomy of animals and humans, and the history of ancient Egypt and the Civil War.

I am awkwardly throwing out my nets as my legs unsteadily waver. My head spinning from these words, time has shattered! A rift in time has split the world and I straddle it, desperately trying to hold on to the reality before, but pulled by the after.

"What? How? What happened?"

"Sir, here is my card. On the back, I wrote the number for the coroner. You need to call him. He can answer much more than I can."

I took the card and stared blankly.

"Are you okay?"

What kind of question is this? How could I possibly be okay? My head shakes a yes/no/go away combination which the officers understood. They left.

I helped Donna inside, and we sat on the edge of the bed,

wailing.

"Coroner's office."
"Hi, this is David Lloyd, and I was told to call you."
Yes, I am being awkward. I could not say my child's name yet.
"Are you Caleb Lloyd's father?"
"Yes."
"I am sorry for your loss. He was found yesterday morning ..."
I won't go into details of this conversation. The Coroner assured me he died by suicide without saying it. He would not say more with the case still open.
"Are you sure it is him?"
"We found his wallet with driver's license."
"What about his phone? He goes nowhere without it."
"It has not been found."
This was enough for me to think this was not my son. When we went on a cruise, Caleb had his phone even though he knew there was no service. Absolutely, he always had his phone. I also knew the coroner is not likely to identify a person without a solid belief.

I called Will, my dear friend and pastor.
"I wonder why he did not answer." I was not expecting an answer and was not really questioning.
"I heard him say this morning he would leave after the second service for Florida."
"Oh, right, he is speaking at the youth gathering in Panama City this week."

I tried calling an elder, then another. No answers. Donna then called a dear friend who lives around the corner. We needed someone we trusted here now! Still no answer.

Our friend returned the call a minute later. "What's wrong?"
  "Ca... leb is, he is ... dead.", barely choking out the words through tears and sobs.
  "We tried to call Will, and..." We listed off who we called to get someone here.
  "I am across the county, but I am coming now!"

Alone is not the correct word, but English does not have a word for the feeling when a piece of your spirit rips away and immediately makes its absence known. My son and I had conjoined souls, a symbiosis. We shared humor, a love for science, and tactical strategy. Now he is gone. Though it felt like hours, we anxiously sat alone for thirty minutes.

2

Blind Faith

> *To one who has faith, no explanation is*
> *necessary. To one without faith, no explanation*
> *is possible. - Thomas Aquinas*

1977

"Shh!"

Mom rarely handed out a warning in church. I often made comments to those sitting near me in the middle of service, loosely related to the message, though usually it was a snarky remark. Some things have not changed. These actions are in perspective after self-diagnosing severe Attention Deficit Hyper-activity Disorder (ADHD), but my younger self did not understand the inability to have self-control.

"David, front pew!"

My father pastored churches from when I was tiny and did not hesitate to scold any of his boys in the middle of the service, nor give a second thought to using our escapades as a sermon illustration. My embarrassment did not seem to slow them down and the audience snickering added fuel for next time. From the week after my birth, they had me attend every Sunday morning, Sunday night, a mid-week service, monthly Saturday night gospel music sing-a-longs, and many revivals. Most of the extended family are devout Christians as well, so as far back as I can recall, praying, singing hymns, and reading the Bible impressed an identity as a Christian. "I cut my teeth on the pews" is a saying picked up somewhere, but is likely the truth. The long wooden benches were as much a part of the church experience as steeples are to others. All the preschool kids slept on them and under them. A quick calculation has me meeting in church over 6,800 hours before adulthood.

Santa Claus

One Sunday night when I was seven-years-old, my father showed a film about Hell in church. Previously hearing it was a place full of fire where bad people go when they die, this presentation made the place feel real. It terrified me such a place existed. Over forty years later, the fear from this experience still has an affect. Of all the horror movies watched in my twenties, none has scared me like this. At the end, I practically ran to the altar to avoid Hell. A Sunday School teacher helped me with my words, repeating after him.

"Jesus, I know I am a sinner, and I ask for Your forgiveness.

I believe You died for my sins and rose from the dead. I turn from my sins and invite You to come into my heart and life. I want to trust and follow You as my Lord and Savior."

Any child would have done just about anything after seeing this film to avoid going to Hell. The prayer comes from the ideas in Romans 10 and not some magical spell. The spoken prayer had truth, but how many seven-year-olds really understand these complex topics? How many are being scared into reciting the plan of salvation in fear?

Attempting to not sin created miserable failures. Any wrongs caused the fear of Hell to creep in, followed by promises to God it would not happen again. No child can maintain this level of perfection. The punishment from my parents would not be as bad as the mental lashing I would give myself for tempting Hell. The church often emphasized the likelihood of not confessing some sin and thus risking going there. This viewpoint is a constant reminder of one's mortality. At the same time, I am agonizing over whether the little lie just told to my sibling would send me to Hell if I get hit by a car while crossing the street.

Following the lead of those around exemplifying faith, trust in God occurred naturally. It becomes part of your psyche and imprinted within you. When a child learns through those they trust about Santa Claus, they will believe he is real and he will only bring gifts if you are obedient. The only proof the child has is their trust in the person saying it and the reaffirming messages they get from others. Children do not fully understand life and death, morality, sacrifice, or even love. They take their cues from those they trust.

There is a difference when people go through a trial of faith or a conversion to faith at an older age. They have some pivotal moment where their mind changes to believing in God, trusting in Jesus, and choosing His way of life. When a child grows up in a church full of people they trust, faith in God comes too easy. The ADHD caused a lack of focus, leading to finding things to do during the frequent sermons. Reading whatever material around the seat became the object of the distraction, and that was often a Bible. This Biblical study led to knowing more about God and the Bible before starting grade school than most learn about American history in school. Not only did I always believe God existed, He seemed more real than George Washington. The church taught that Jesus is closer than a brother and I had three brothers. Going to the altar that Sunday night had a significant impact, but it did not feel like a dramatic conversion.

The very act of questioning my faith would have been at odds with my identity, and thus the idea never occurred to me. It was a given; a part of life like which side of the road to drive a car. A child can see the lines on the roadway and the thousands of cars staying in the lines. Faith at this age is the same as knowing how roads work. Kids do not understand how to drive a car, but they will tell you they do because they think they truly can. They know to get in the car and follow the path to their destination. I saw people who trusted and relied upon their faith. Every day someone would talk about it. They would also tell about answers to prayer they had received. This combination reinforced my belief in letting faith take me to my goal, Heaven. Many Christians push blind faith. They believe in faith like a child and mistakenly translated by some as believing in Jesus like children believe in Santa Claus. Pastors and leaders with an abundance of faith tell those questioning to have more faith; pray for more faith.

When faced with problems, the church taught to pray, believe, and our loving God would take care of us. The churches my father pastored were small and had little money. He moved us one thousand miles to pastor his second church, the first I can remember. Before we came, it had dropped in attendance to three old ladies and the church had not paid $800 of its utilities. With inflation this equates to $3500 in 2020! My parents had little money and the ladies only had social security income. Pregnant with their fourth son during the move, the church provided its basement, designed to be fellowship hall and classrooms, as living accommodations. A few months later we visited Grandma for Christmas, and upon returning, found our living space flooded. Since my parents grew up in south Florida, they did not know pipes would freeze in the Indiana winter.

Mom would buy all our clothes from thrift stores, consignment shops, and garage sales. Dad had to convince the utility companies to keep services on until we eventually paid the balances. People in the community helped with cleaning, repairs, and babysitting while Mom delivered my brother. Being young, all this adversity never concerned me because those around me had no great concern. Faith can be instilled without ever discussing it. A couple of years later, the church was running about eighty regularly and my father moved to another small church.

We lived across the street from the church in an economically depressed area of town nicknamed The Projects. However, this was not typical government project housing.

The military quickly built a bunch of cheap multi-family homes during WWII. They were mostly cheap rentals and HUD assisted housing. At nearly ten-years-old we moved, and then my grandmother bought us a television. My parents were not happy, but us kids thought it was a treat! I watched cartoons and whatever shows on the four channels received. Then, someone broke into the house and stole it not too long after. Drug dealers lurked everywhere and the kids in the neighborhood did not like me, so I spent most of my free time in the public library reading. Again, this church had no rich benefactors. A simple group of people, helping other people in whatever way they could, and praising God for what they had. If you had a leak, or your car needed repair, or you could not pay a bill, someone was there to help. Then everyone thanked God for providing. Once the church was running about eighty in attendance, Dad moved us again.

This time we moved one thousand miles back to where I was born, and we stayed in one of my great-grandfather's houses two doors down from his small commercial fish business. He and my great-grandmother lived at the business along with my grandmother, who provided care for them. The four of us boys had to sleep in one small bedroom together, but I loved living near my relatives. After school I worked alongside my uncles and great-grandfather packing fish, making boxes, and cleaning around the dock. This instilled powerful family and work values. Despite the hard work, the time spent with family and not dodging bullies felt like a rest. Unfortunately, it lasted only two years.

Once again we moved and were over five hundred miles from family. Although in a completely different city than before, our accommodations once more existed inside a small church's fellowship hall and its classrooms. The pantry rarely had much food, but we never missed a meal. Sometimes we would not know how the next meal would happen, but someone would unexpectedly give a little money. At least once, an unknown person placed several grocery bags of food on the porch when we had no food for a meal. I was very thankful for these just-in-time answers to prayer. Seeing these timely acts of compassion taken by people who would have not known of our immediate need, my faith in a kind and loving God strengthened. It seemed to me He ensured provisions, but not normally ahead of time.

Angels Watch

1984

Mom and Dad moved several times in 1984-1985. Attending three different high schools my freshman year at schools which did not all have the same classes, I struggled and barely passed. Not knowing any classmates made the already challenging year much more difficult. The last move put the family back in The Projects we left when I was twelve. The few friends from back then moved away or held back a year, thus they were not in the ninth grade.

Not much had changed since moving away, except the streets had more drugs and prostitution, or I recognized it more. The church had lost most of its attendees.

I hung out with a few people occasionally, but with no consistent friends as a distraction, I buried myself in choir, marching band, theater, and computers.

1988

Dad would sometimes ask about career choices, however thoughts of college did not occur until the senior year of high school. Neither the school nor my parents pushed it as an option. This small town had no future for me, and I wanted to do more. There was no way they could help pay for it, but my SAT and ACT scores opened the possibility of free tuition to some excellent schools. After touring Ball State University, about three hours from home, I paid the housing deposit and some fees. The state headquarters for our church then announced a trip to Lee College for a long weekend. Some acquaintances had registered and it would be fun to get away. While visiting, God told me to go to Lee. As a liberal arts school sponsored by the church, they forced all students to get at least a minor in religion, but had a great reputation for music and teaching, which is what I wanted. Applying on faith the funds for an expensive private school would come, Lee awarded a full tuition with room and board scholarship. Along with a confirmation in my heart, I took this as a sign that God wanted me to attend Lee.

Having a full scholarship does not mean coasting into college. Money is stll required to make the six-hour trip, maintain an eleven-year-old car, buy books and supplies for classes, and living expenses on campus. It would also be nice to have some extra for an occasional outing with friends. With an overloaded schedule of band, theater, and newspaper class, I found time to work as construction help.

After graduation, I worked at a fast-food restaurant as a second job. The summer of 1988 was abnormally hot for southern Indiana. Many days were in the high 90s and a lot in the 100s, but every day had a heat index over 100 degrees. Scheduled to work the grill flipping burgers on one of the hottest days, I asked my mother to let me drive her car to work since her air conditioner worked and mine did not. Unfortunately, the air conditioner compressor for the grill area had stopped working. It was boiling hot working conditions.

Hours standing over at the grill made me very ill. They reluctantly agreed to allow me to leave early. While driving home on the four lane divided highway, the illness worsened. The tires on the left side went off the road a few inches. My old car had a very loose power steering and would take a lot of movement to turn. Instinctively yanking the steering to the right as my car needed, her car shot straight across the highway, hit the drop-off on the other side, flipped, bounced up, flipped, bounced up and landed upside-down on a military fence running beside the roadway, then landed on the ground.

The fence bent the top of the car down in a 'V' shape. The roof reached within two inches of the headrest. This should have crushed my head or thrown me out the windshield! During the first flip, I arced over the headrest and landed in the back seat. The remainder of the ride I bounced in the back seat. Broken glass became embedded in my hands and back. If I had been wearing my seat belt, no doubt I would have died.

The radio had been on and when the car came to rest, it was playing "Angels" by Amy Grant.

"Angels watching over me, every step I take.
Angels watching over me!
God only knows the times my life
was threatened just today.
A reckless car ran out of gas before it ran my way.
Near misses all around me, accidents unknown
Though I never see with human eyes
the hands that lead me home"

I stepped out of the back door. I later learned it was the only one operational. As I walked up the grassy shoulder towards the road, a man jumped out of his car yelling at me. Living in the projects and working with many commercial fishermen, rough language flew by with frequency. This guy took the prize for the longest sentences formed with nothing but curse words. Concerned for my health, he wanted me to sit down. To calm him, I rested on the grass, staring at the damage.

Someone went across the highway and phoned for help. The ambulance transported me to the hospital to make sure I had no internal injuries.

The nurse called my parents, but they were at church. Whoever answered the church's phone thought the hospital said I had head and shoulder lacerations, and missed them saying they would discharge me soon. This message worried my mother, so she grabbed a family friend to drive her to see me.

The route to the hospital took my mom by the accident scene and she saw her car; corners and hood crumpled, top crushed, and axles severely bent. Anyone would think the occupants could not survive. They drove up to the hospital to find me sitting on the sidewalk. I had a couple of bandages and a sprained ankle.

Were angels watching over me? Why did I not have a seatbelt on? How did I pass over the seat without my long legs catching on the steering wheel, and how did my long torso clear the seat? This was a compact car and I am six feet tall. It threw me onto mounds of broken glass several times and my cuts were superficial.

My ignorance about heat exhaustion led to the accident. I should have sat in a cool area, drank water, and waited for my body temperature to return to normal. I also believe no one arrives at the Pearly Gates by surprise. God has numbered our days (Job 14, Psalm 31, and 139). If it was my time to die, I would have. Whether He used sheer luck, angels, or divine providence is not material. This was yet another tremendous boost of faith for me.

Pray, believe, and God provides. This is what I saw growing up, and it was all I knew. I needed to learn a lot about everything, including God, but what could be so bad if He was on my side? Blindly trusting in what you believe is like being on the water with no life jacket. Everything is comfortable when it all goes smoothly, but when life throws something at you, you may want something to hold on to tight.

3
Faith through Jobs

*Belief is a wise wager. Granted that faith cannot
be proved, what harm will come to you if you
gamble on its truth and it proves false? If you
gain, you gain all; if you lose, you lose nothing.
Wager, then, without hesitation, that He exists.*

- Blaise Pascal

After arriving at Lee College, I signed up for a work-study
assignment, a financial hardship based job. God blessed me
with an immediate job and kept me in one throughout my
stay at Lee. I worked in the cafeteria, student center cafe,
security, and maintenance (mostly changing light bulbs all
over campus). Having helped with lights and sound in high
school, they transferred me to the main auditorium to help
set up and run their sound systems. This in-depth teaching
matched with music knowledge has allowed me to manage

and operate the sound for churches big and small for many years.

At fourteen-years-old, I begged my mother for a TRS-80 Color Computer II with 16K of RAM for Christmas. It went on sale for $79, so she bought it as a gift for the family. Even at this cheap price, it pushed their budget. Having enjoyed developing software for countless hours, I took a computer programming elective class to get more formalized training. The course used a programming language very unlike the TRS-80's, thus making it more intriguing. While the class fascinated me, the idea of working in a large corporation and helping them write boring programs held no interest at all. I planned on pursuing higher-level math theory and teaching.

A few days before the final exam, the professor asked me to stay a few minutes after class.

"Are you going to take any more computer science classes?"

"I am not sure, but I really enjoyed this one."

"I am adding more classes and need an administrator for this lab. It is a ten hours a week, minimum wage position."

"I have a work-study job doing sound in the auditorium."

"This is paid by a grant, not work-study. You can do both jobs."

"I'd love to do it, but I'm not sure how though."

"I will teach you."

I could double my hours and do two things I loved, running sound and playing with technology. Running the lab allowed me to learn the Unix operating system, learn how to put together computer systems and hardware, and pushed me into studying yet another computer language. I immersed myself into learning all of this while taking a full academic load in Mathematics. I loved it! Somehow, I knew this was the future and I would make it my future.

Who Needs Sleep?

June 1991

After getting married, more on that later, I lost my job delivering food because my car continually broke down. My bride needed me to support her properly. I prayed for the right job which allowed us both to still attend school. A friend working at a facility for juvenile delinquent boys recommended me for a night shift position ensuring the safety of the residents and calling the sheriff if one ran away. While not glamorous, none could match its perfection. The tasks required me to sit in a central location and occupy myself, only getting up once an hour for a bed check. I did homework for about three hours, studied for tests, and then had several hours to immerse into learning computer programming.

What a sight! Each night I would arrive a few minutes early to clock-in and then move in for the night. A stack of books at least a foot deep would be brought in. The math and programming book thickness alone weighed a lot. Additionally, I hauled in a large computer case and amber monitor along with cables.

The residents would laugh, but I did not care. I had a dream, and worked on my plan to make it come true. I kept my work study assignment doing audio, continued to run the Unix lab, and worked this full-time job which allowed me to take a full class load. What could be a better answer to my prayer?

Jackpot

1993

A few weeks before graduating, I searched for one of the Information Technology professors. The work study and lab jobs would stop at graduation, and he often had leads for entry-level technology jobs. He was talking to another graduating student.

"I just put in my resignation. They asked if I knew a student who could do technical support."

"Okay, I'll ask around. Someone would love a job at a software company."

I had walked up at the perfect moment.

"Hey, where is this? I am looking for something in software."

"It is off Michigan Avenue. They do church management software, but they want someone to handle technical support phone calls."

"Sounds great!"

I went to the interview and expressed my interest in programming and outlined what I knew. In retrospect, my knowledge was close to nothing, but I did not know it.

"We do not have a position as a developer right now. If one opens up, we can talk about it. Are you interested in technical support?"

"Yes!"

"When can you start?"

"A week from Monday, the day after I graduate."

"Perfect. We'll see you then."

I prayed for a specific type of job, visited a professor to begin a search, and arrived at just the perfect time to get a position with a software company only four miles from my apartment! Six months later, I moved from technical support to programming and made twice the amount as the year before. This is how prayer works!

Weirdness Pays

April 1996

Several years later, the company sold to a corporation with a different focus. Knowing my position would not last long, I immediately found a job as a software engineer for a major airline. Donna expressed her concerns moving all our belongings and switching health insurance at eight months pregnant. I assured her that the moving company would handle our stuff and the insurance covered the pregnancy. She went to the doctor and spent $5 on a copay to prove, once again, she was pregnant. They covered everything else.

After working for the airline a while, I left to work at a smaller company for several more years. Continual pay raises and increased responsibility kept the bills paid and the jobs interesting. I have a weird brain wired differently than anyone I know. I easily see how to solve abstract problems and achieve goals in large computer systems. This ability commonly exists in people who have Aphantasia. People around me see this skill and reward it. My job situation could not be going much better.

I had faith He would provide, and why would He not? In Philippians 4 He says, "But my God shall supply all your need according to his riches in glory by Christ Jesus." In Matthew 6 He says, "Therefore take no thought, saying, What shall we eat? or, What shall we drink? or, Wherewithal shall we be clothed? ...for your heavenly Father knoweth that ye have need of all these things." The Bible makes it clear here and in many other verses that God provides all our needs. In modern first world countries, our needs are often provided through a job.

I leaned on God's provision to keep my faith bolstered. If my personal life matched half this goodness, I'd be the happiest person alive.

Pay It Forward

November 2001

In my eighth year of software development, while at a different company, they laid off all the regular employees shortly after 9/11. Not a lot of businesses had open positions. Companies, along with the rest of the country, feared the threat of more attacks and the dot-com boom had busted. After several months of being out of work, the savings disappeared and bills piled up. I wondered how these would get paid. Searching for a job constantly, going to networking events, and asking friends, I would take any position. Then the mortgage payment due date neared with no money to cover it. During this time I played trombone in the church orchestra. The next Sunday morning, a member of the orchestra walked up to me.

"I feel like I'm supposed to give you this."

He handed over a check for the amount required. Overwhelmed and filled with gratitude, I promised to pay it back.

"No! Pay it forward instead."

This check paid the mortgage and food; exactly what I needed for that time. It allowed me to go a little longer, and able to land the next job.

I use the term "job" loosely here. It was an oral contract with no paperwork. Our handshake agreement had me show up and do what they required, and get paid at the end of the week. Every Friday for over four years, I would pick up my check. We had a half-joking banter we would do.

"See you on Monday."

"Only if you will pay me next Friday."

"Deal."

It was not long after starting this work I repaid my debt by paying it forward.

The last six months of this job had me working as a subcontractor for another corporation. A Fortune 100 company then bought it. They offered most employees a job, but this did not apply to me as a subcontractor. The job search began again. Caleb had joined the Cub Scouts, and they cajoled me into the role of Cub Master. The Assistant Cub Master would stop on his way home from work to pick up awards, trophies, badges, and other materials from the Scout Shop.

However, he often showed up very late. One important award day that the scouts received achievements and awards, they held nothing in hand until the end because he arrived late. Afterwards he apologized profusely, explained what happened at work, and then got to his point.

"You do websites, right?"

"Yes, I do websites but specialize in finance-based web applications."

"Have I got a job for you!"

Who's the Boss

April 2006

The Assistant Cub Master had created a simple, internal use only, website for his employer, a very large multinational company. Although not a part of his duties, they loved his idea for tracking the sales process. They thought it would take two, possibly three, months of work for me to finish what he had started. Far exceeding their expectations, I turned their idea into a much larger concept and took on some other projects from them as well. This required me to hire a team of subcontractors to help. We all worked together so well we have continued to be great friends over the years since. I provided a full-time team to this company for over seven years.

Professionally as a software developer, I have never been happier or more fulfilled than the seven years managing this contract. Part of it was the fantastic team and the resulting working environment we created. Being able to do this from home opened up many possibilities from spending more time with family, taking on other smaller contracts, and doing more with the scouts and our local church. Sure, I worked more than ever, but I also spent more time with family and community than ever.

2013

Shortly before this contract was over, the other little side jobs had also stopped. I spent months looking for another contract, a regular job, or even small side work of any kind. This continued on until my bank account dropped to a dangerously low level. With only two weeks of money left before bills would stop being paid, a company made a job offer which I accepted.

This pattern of getting jobs miraculously, and coming just in time, stayed with me. The Bible is full of stories of God showing up just in time, but not a lot before. The examples show God pushing His people to rely upon him. Not everyone has the same story. Some people end up homeless after losing a job. I am not critiquing their faith, nor saying God provides the lifestyle we want. However, this string of circumstances in my life built my faith, and I was going to need a lot of it.

David A. Lloyd

4
Faith through Hard Times

The deepest definition of youth is life as yet
untouched by tragedy. - Alfred North
Whitehead

August 1990

Every year I attended a Back-To-School concert at Lee
College with friends. The start of the third year was no
exception, a packed auditorium had me and a friend searching
for seats. I found some next to a girl I met earlier. I never
believed in love at first sight, but this was the opposite. We
had a strong dislike toward each other after the first few
minutes. She thought I acted arrogant, and I thought she

behaved snobbish, though I thought my friend might like her.
"There are some seats!" I motioned for him to go first. We
were close to her, and I was trying to be discreet.
"Oh, good. Go right ahead."
"No, you should."
He shook his head, so I sat. Her demeanor changed, and it
became obvious I had not been discreet. Later, I caught up
with my friend.
"Hey! I think you should ask Donna out. I tried to get you to
sit by her."
"What? No, she is not my type."

A few weeks into the semester, Donna and I ran into each
other on a Saturday morning in the Student Center. We
started a conversation. As it progressed, we let our guard
drop, and we did not stop talking. I did not journal or record
this day, but I wish I did! Her dorm was the two floors over
the Student Center.
"How is living over all this activity?"
"It's OK and a convenient location. The water pressure is
pretty bad though."
"Yeah, my dorm's water pressure is terrible too."
"Did you get matched with a nice roommate?"
"I am rooming with another girl from the Home."
I knew she had lived at a home for kids and had driven by
this place before, but knew only a little about it. We chatted
about the home and what led her there. The conversation
continued deep but moved through my past and what we each
wished for each of our futures.
Fourteen or more hours we talked. A "normal" American

couple likely has light banter on the first several dates. They talk a little on quite a few more dates. Couples take time before they open their emotions and come to know each other. We shoved something like twelve dates' worth of communication on this first day together.

We walked most of the day. Breaks from walking occurred on various benches, a porch swing, and steps of classroom buildings. Late in the night, we sat on a bench outside her dorm. In the moonlight, we kissed. No couple has ever been more different, had as little in common, or been more in love.

We had more walking talks several times a day. This allowed us to learn each other rapidly. We were inseparable. Long discussions on family life and child-rearing, Christian beliefs, and where we used to work. These in-depth talks also had conversations about children and how many we wanted. I believe God placed the number four in my thoughts, and I said it. This was a bit of a surprise to me, and then I learned she desired five or more.

So how does one go from a strong dislike to a strong like in a day? While infatuation at first sight is understandable, and lust at first sight exists, I place little stock into seeing a potential mate and having a deep love at first sight. I never said she was not attractive. Donna is very nice looking. I cannot appreciate good looks if the inside is bad. Once I knew more of who she is, I learned her attitude differed from what I assumed and her inner beauty complemented her outer beauty.

Genuine love comes by knowing someone. Not satisfied with just knowing, I wanted to know more. After 30 years, I am still enthralled to learn things about Donna. It also seems crazy that there would still be things unknown. In 2017, I wanted to spend more quality time together as a family in a simpler setting, getting exercise and teaching our daughter outdoor skills. Donna and I had camped often when we were young, but many years had gone by with no trips. I thought her fondness for camping had passed. She did not just agree, but dove into it with passion.

Despite all we have in common on a spiritual level, we have learned we have little else in common.

A week after this first date, the children's home sent a van to pick up Donna. They wanted her to come back and "refile paperwork". It seemed odd to us. We consulted with the President of the college and he advised it would be best to just go along. He knew these individuals and so did Donna, so she went.

The people from the home put her on a plane and sent Donna to her parents against her will. I was not happy with this outcome and it felt wrong. She was eighteen, and I thought they effectively kidnapped her. It certainly appeared like they had coerced her to abandon school.

Our rocky road begins. We talked by phone many hours a week for months. I worked extra jobs to pay the long-distance bill. We knew somehow this would work out to be together.

I drove seven hours to visit her and meet her family at Thanksgiving dinner. Donna and I met at her friend's house, who graciously allowed me to stay for the holiday. When it came time to go to dinner, Donna's parents told her I was not invited. The friends where I was staying set an extra place for me and treated me as family.

Good Luck!

Donna returned to Lee the following semester, and again rarely separated, she accepted my proposal a couple of months later. We received some strange reactions from most everyone we knew.

"You guys won't stay married a year."

"Yes, we will. Why do you think we won't?"

My parents' reaction bothered me most. I know this conversation can go several ways and braced for the shock and disappointment to come.

"Hi Mom, Donna's pregnant." I answered some questions, but what came after my next announcement surprised me.

"We're getting married in May."

"What? You know you don't have to do that! You don't have to marry that girl."

I thought she would expect me to wed her and hoped it would elate her.

"I want to marry Donna."

My reading of the Bible, my understanding, had me already married in the eyes of God. To walk away would be adultery.

With friends doubting, my parents trying to talk me out of the wedding, and her family still estranged, we eloped. We told many friends when and where it was happening. The planned meeting occurred at the courthouse one hour after my last final exam of the year, nine months after we met. We had two close friends show up with us. Perfect! A best man and maid of honor to witness, along with Alex growing in Donna's womb. The judge remained seated, eating his lunch.

"Come in. So you kids want to get married?" The tone of his voice suggested he expected to sign our divorce papers in a few months.

"Yes, sir." I hand him our paperwork.

"Do you have your own vows?" He takes another big bite of his sandwich.

"No."

He looks at the marriage license. "Do you, David Lloyd, take Donna as your lawfully wedded wife to have and to hold, from this day forward, for better, for worse, for richer, for poorer, in sickness and in health, to love and to cherish, till death do you part?" He somehow swallowed his food while saying this. He sounds like he has done this ten thousand times. The judge then took yet another bite of his sandwich.

"I do."

"Do you, Donna Ritchey, take David as your lawfully wedded husband to have and to hold, from this day forward, for better, for worse, for richer, for poorer, in sickness and in health, to love and to cherish, till death do you part?"

"I do."

"David, you may put the ring on Donna's fourth finger."

I nervously reach to comply exactly as he said and count in my mind, skipping the thumb: one, two, three, and I go to place it on her pinky finger!

The judge corrects me. "That's the fifth finger."

"Donna, you may put the ring on David's fourth finger."

"I now pronounce you man and wife." The judge wipes his mouth, cleans his hands, and shakes our hands. "Good luck!"

Buried Grief

I took the day off work and arrived the following day.

"Why did you call in yesterday?"

"I got married!", flashing my ring.

"What! Why aren't you on a honeymoon?"

"I don't have money for that!"

I walked away laughing, not caring I had to work. Happiness flooded me.

"David! What was that noise?"

"It was just thunder.", still mostly asleep.

BOOM. The apartment shook. Donna woke me again.

"That's not thunder!"

Having looked out the front door, then the back seeing nothing, suddenly several teens ran up to the apartment building's wall and a car appeared backing out of the side of the building! A teen drove into my neighbor's home about ten feet from my bed, on the other side of the wall. The boys jumped in and sped off, dripping fluids and dropping bricks from the wall. The police showed up forty-five minutes later.

"What happened?"

"Some kids took a car, drove it through this grassy area between these buildings, and ran into the apartment. They backed up and then hit the wall again. Then they backed up and left."

"Which way did they go?"

"Follow the yellow brick road!", pointing at the trail of

fluids and bricks. I am naturally sarcastic and even more so at 3 a.m. after waiting so long for a response. This neighborhood was the cheapest in town for a reason, but it is all I could afford. We trusted in God for our safety and to make rent.

Happy to start our family and wanting a large family, less than two months after marriage, we suffered our first miscarriage. We had discussed names and Alex was one we were debating. It stuck. We grieved about losing Alex. How can a life so easily conceived, with no effort or wish, be loved so deeply while unseen? Love for a child, both the born and unborn, does not subside.

I did not talk about the miscarriage. I heard no one talk about our pregnancy or loss. It did not occur to me then how insulated we are from child loss. People do not just not want to talk about it, they go out of their way to avoid it. The first several people I told visibly clenched up, lips tight. I pushed my emotions down, deep down. I denied it affected me for over fifteen years! Many friends had children shortly after and those children were, and still are, reminders of what could have been.

Nearly half of all who experience a miscarriage feel they have done something wrong. This also includes their partners. In reality, almost all causes are factors beyond control; most commonly are genetic abnormalities in the embryo. These misconceptions cause direct psychological harm. Pregnancy loss studies found it can traumatize and cause guilt. The worst symptom, because it is so preventable, comes from the isolation and difficulty disclosing. Talking with others about your miscarriages helps those currently going through the same grief. Acknowledging the pain, giving hope without platitudes, and holding space while your friend walks through their hurt, brings comfort to most who are suffering.

I buried myself in work. This has an odd logic, burying myself working to deal with burying my child. Besides a full academic load, I worked night shift forty hours a week, also in the auditorium and Unix Lab, and continued teaching myself programming. This left time for four hours of sleep and a few minutes with my bride except on my days off. I have since learned my primary coping mechanism when dealing with emotional trauma is staying extremely busy.

Eating is another way I cope with stress or trauma. Being without money does not stop access to food as the job always had snacks, hotdogs, or something around. I would bring colas and cookies when I could afford it, or use the provided coffee and sugar non-stop to stay awake. This lifestyle caused a weight gain of forty-five pounds in seven months.

The grief underlay all I did, but it was never very prominent. I knew miscarriages were common and had faith God would allow us to have more children. We were young and hopeful our family would continue to grow.

Survival

Shortly after the miscarriage, Donna fell in the shower and re-injured her previously broken tailbone, leaving her mostly confined to bed for months. While I worked all night and went to school all day, the friend who was our bridesmaid helped nurse Donna's wound. When Donna was also working, we made a little money, but paying for college for both of us took all of it after living expenses. With Donna laid up in bed, we were not making enough to pay bills. I applied for assistance with the Department of Human Services. They said our income exceeded the limit by $5 a week, but would be eligible if Donna worked. If she could work, I would not have needed help!

We scraped by on pennies, even fighting the bank over thirty-six cents removed from our account in error. Five or more nights a week, dinner came from dented tuna cans, dried noodles, and mac-and-cheese. Once we could afford otherwise, I refused to eat tuna for many years. One night Donna ruined dinner by leaving it on the stove too long. I had to leave for work soon and we had nothing else to eat. The only money in our bank account matched what we needed for bills with no extra. Our wallets were empty. After scrounging in couch cushions and the car console, we barely found the amount needed for two burgers and fries at a super-cheap fast-food place nearby.

Another time, I walked in the door to Donna on the phone.

"Some close friends I have not seen in years are going to be outside Nashville tonight. They are leaving in the morning. I'd like to go see them but I don't think we can afford it and you have a Calculus final at 9:00 a.m. in the morning."

"I am ready for the final. It is a little over two hours' drive if we cut north, ignoring the interstate. It saves an hour. We should go. Who knows when your next opportunity will be."

"I'll call them back and let them know."

Taking the alternate route led us through the mountains of

central Tennessee and very little civilization. I noticed along the way the car was acting weird and the lights flickered unsteadily. I suspected the alternator was going out.

We met her friends at a restaurant and talked until it was late.

"It was nice to meet you! We really need to get home." We walked to the car and it would not crank.

"Could you jump start my car? I have cables."

"Sure." After letting it charge up a while and then cranking my car, "Are you sure you'll make it home?"

"I think so." We changed the route home to use the interstate, so if the car had more problems, maybe someone might offer to help. After a few miles Donna became distressed.

"David, the lights are not shining as bright as they were."

I pulled into a rest area, disconnected the battery, and carried it and the charger to the building. We sat on the floor for a long time near the only convenient electrical out waiting for a charge. Completed, I reconnected the battery and traveled toward home. I stopped several more times to charge. Almost dawn and still thirty miles from home, the car stopped limping home. Thankfully, I coaxed it to an exit and parked on its shoulder. The only business nearby was a credit union which did not open for another hour. There were no payphones around, and affordable cellphones would not exist for several years. We waited once more. When the credit union opened, I called a friend who brought us home. Well, I did not go home. He dropped me off at the school to take my Calculus final exam, which had already started an hour earlier. The three-hour trip home via the interstate took over ten hours.

With no sleep and only half the time remaining to take the exam, I nevertheless made a good grade on the exam and the class.

Even though I judge myself now for working too much, I doubt we would have survived on anything less. I worked the night shift for two years while working my campus jobs and carrying a full academic load. Sleeping only a few hours in the afternoon ruined my memories of those years. I recall very little of my early years of marriage, and the time spent with close friends living within walking distance. How we made it through those years without ending up homeless or divorced is a miracle. Maybe it was blind stubbornness.

5
Faith through Infertility

Hope is a good breakfast, but it is a bad supper.
— *Francis Bacon*

1992

"They offered me a position working day shift weekends! It is Friday through Sunday, twelve-hour shifts. I can start sleeping at night now. It comes with a small pay raise too."

"Oh? That's awesome! You have not been sleeping well and it will be nice to have you home."

Donna had been trying to nudge me towards having a child again ever since we lost Alex. Because we both attended school and my job had been tenuous, I thought it best to delay having children. I do not recall if it was the same day I found out I had a better work schedule, but not much time passed before Donna's nudging became more vocal.

"I would like to try for a baby."

"Wouldn't it be easier to do after you graduated?"

"That's waiting two more years! I think we can handle this now. Do you?"

"Okay, it will be at least ten months before a baby arrives. We are doing well and I will graduate soon. Hopefully, I can find something more of a career." Donna had returned to working, and we felt less financial stress. My making a little more money helped too.

A few months into trying to have a child, the software company hired me for technical support and things were going well with our life. This just added fuel to our desire to have children. In the beginning, attempting for a kid moves like a calm pursuit. While actively changing our routine, we maintained hope and tried to not stress. After nine months of earnestly trying, we wondered why it did not happen yet.

"David, get in here! It's time!"

Maybe the timing is off. Are we trying too often? Not enough? We tried using a calendar. Then we added a basal thermometer and scheduled our intimacy.

"You need to come home for lunch!"

Over a year goes by and we are getting stressed. Donna is crying most days.

"Why is nothing working?"

"I have no idea. Maybe we should see a specialist and get checked out."

We discovered Donna has polycystic ovary syndrome (PCOS). The doctor said her ovaries did not release eggs regularly, but we might have success continuing to try. We doubled down on the temperature checks and scheduling intimacy. Many more months go by and life now resembles the opposite of romantic or calm. We continued normal lives outside the home; working and joking with co-workers, often visiting with friends, and had a good time with them. However, inside the home was a train-wreck. Being at work and seeing friends distracted us from the problems at home.

"You should divorce me! You deserve to have a family and my body won't cooperate."

"No, I am not getting a divorce. I love you and, even if we do not have kids, I still want to be with you."

The stress helped nothing and probably made it worse, but we do not possess infinite patience. Patiently handling the catapult into starting a family early and having it ripped away, we waited for the right time to resume and patiently bided time for almost two years while trying to become pregnant.

"Why could I get pregnant so easily before? God hates me!"

"I don't know why, but I don't think God hates you."

"Then why am I being punished?"

These are tough questions, and I did not have answers. I let Donna know daily how much I love her, and we will make it through this. Faith remained in the four kids God placed in my heart. Constantly reminding myself and Donna that our schedule is not God's schedule, I prayed every day for our situation. While it seemed like I was calm, I did not know why we should have to wait. Knowing children will come and waiting for it are two very different things. God promised Abraham a child, and he waited twenty-five years for Isaac!

Over three years passed with us trying to get pregnant, and the waiting was excruciating. Over two years of daily temperature checks, calendar marking, intimacy scheduling, and emotional roller-coasters. I prayed over 1,000 days, believing God would send our child. Every time I took the dog outside or for a walk, I prayed. I begged. I cried.

Instead of becoming pregnant, our friends did. It was not just one or two friends, but every one of our close couples and most of the other couples with which we would hang out. We wanted to be with our friends and loved holding their children, but doing so also stabbed us as a reminder of what we were trying so desperately to have. Then two of my brothers had kids, and one of Donna's sisters had two kids. This was becoming ridiculously difficult.

"The ovaries may not be releasing eggs, or releasing them infrequently, making it difficult to inseminate. You might want to consider using a non-steroidal fertility medicine."

"You're the doctor, I don't know. Is it going to work?" Donna was not just asking. We needed to know it would work. Not only were we concerned about having yet another miscarriage, we did not have funds to cover this procedure and did not want to gamble our savings away.

"It has a high success rate at forcing eggs to be released, but I cannot promise anything."

Understatement

I would not recommend couples to take hormone level

altering drugs unless they were certain of the consequences. They need to be sure of their commitment to each other, desperate for a child, and willing to go through a very rough time. They must be sure they won't kill each other. Having attempted to explain the emotional roller-coaster we rode trying to become pregnant without medication mentioned above, that experience had us basking at a pleasant picnic in the park on a warm, cloudless day compared to our experience with the fertility drug.

Consider a child going through puberty and how they experience the gradual changes in hormone levels. These changes are challenging to handle because these hormones affect brain chemistry and thus emotions, mood, and behavior. Now imagine getting a large influx of these hormones, not over years like in puberty, but in five days. Then another dose the following month, and so on. The package listed "increased irritability" as a possible side-effect; the most understated warning I have ever read.

Each month starts with going to the doctor for a checkup which involves getting, among other tests, a pregnancy test. Each negative test is a depressing punch in the gut. A pill is then taken on five consecutive days. As the hormones ramp up, they throw the emotions everywhere. The hormones also cause pain, tenderness, hot flashes, and other side-effects. As you might imagine, these symptoms do not help with the mood swings. One of the side-effects was weight gain. Maybe it was the pill, or the hormones, or just stress-eating, but the weight came on quick. However it occurred, I gained too. I told Donna my weight gain was sympathetic weight gain even though stress-eating is the likely cause.

Donna had recently graduated with a license for teaching high school business classes. Her prayers during month three and four shifted into asking for a job teaching if she could not get pregnant.

After the fifth month of insanity, Donna went to her doctor appointment.

"You're pregnant!"

We were ecstatic. Happiness flooded our souls, but we were also afraid because of our first pregnancy's miscarriage, so we told only a few very close friends. A few months later, we told the family. In June 1996, almost exactly five years after we lost Alex, Caleb was born.

Hope

I do not know how to convey the extreme joy and happiness at having our long-awaited son! Not only were years of prayers answered in Caleb's birth, the following week a school offered Donna a job! Getting the first teaching position can be slow and difficult, but yet again, our work situation never faltered.

Those years struggling with infertility were grueling. We fought with each other and wrestled with God. I yelled at God, and while more figuratively than literally, shaking my fist at Him. Daily, I cried, begged, and pleaded with Him. I cried with Donna as she appealed her case before Him for several years.

More than just trusting in God, I believed in prayer; prayer which changes things and moves mountains. Knowing He could change our plight, heal the body, and calm the mind, I waited for His timing. I did not know why we had to struggle to have Caleb, but I did not care. He was here!

Now that my son is born, and Donna is teaching, I put those problems behind me. I praised God for my boy and thanked Him for launching Donna's career. Throughout this time, I did not have to pray for my job much. I did, but it did not seem as necessary. It was the only aspect of our life going exceedingly well. Not only did I get a nice pay raise, when I needed to look for work as the company was being bought out, I easily found a great job making much more money. It was odd, the one thing I cared little about was being blessed abundantly. My focus and desire had always been on family. Now that it seemed this was back on track, I took a breath. My faith, while not weakened through this ordeal, was being built.

I was going to need that faith.

Punished Like Rachel

A year after Caleb's birth, we wondered if we might have some difficulty conceiving again. Donna went to her doctor and received confirmation we would have to use fertility medicine, but this time we would need to start at a higher dose. We did not have the money for the treatments and decided, even if we did, we did not want to go through that horrible experience again. In retrospect, I wonder if we should have found a way to make it work.

Seven years we tried!

Seven years of pushing towards a family every month.

During this time, my job is going great. God has provided the timely check between jobs and I am working the handshake 'contract'. Difficulties do not stop me from being with Caleb. He and I enjoyed soccer, scouting, and stargazing together, and his love for astronomy, and all of science, blossomed into a passion.

No, not just seven years trying. Seven years of attempts since deciding to not do those drugs again. We had wanted to build our family since losing Alex thirteen years prior. We actively struggled for eleven years consecutively except while pregnant with Caleb. I know people who went through worse than this, and I hurt for them. If you have not, then I doubt you understand. Everything gets messed up, especially the intimacy with your spouse. You have this shared pain which should make you closer, and it does in some ways, but the physical intimacy becomes an emotional drain. Part of the problem requires scheduling intimacy around fertile times and then dealing with the schedule when she has a migraine or work keeps me from coming home. The other part is the feeling conception won't happen because it has not the other 500 weeks. Each attempt is a reminder of the previous failures. How does this not mess one up psychologically?

So while my job is going great and the surrounding circumstances built my faith, the constant fight to have a big family is tearing at my faith. I knew God interceded on my behalf, helping me get through rough patches between jobs and providing fantastic jobs. I started thinking about the sisters Leah and Rachel in the Bible. In Genesis 29, "And when the LORD saw that Leah was hated, he opened her womb: but Rachel was barren." Was God blessing my job more because we could not have kids naturally?

6

Faith through Adoption

If the mountain won't come to Muhammad,
then Muhammad must go to the mountain. -
old Turkish proverb

2004

I had been around several families who fostered children and some who also adopted ones they fostered. Donna and I had talked for some time to investigate the Foster To Adopt program. We waited until Caleb was old enough to handle himself and tell us if a child was violent toward him. In 2004, we took the classes, background checks, and home studies for approval.

We were not abandoning natural conception, but we were putting the focus on it aside. This allowed more breathing room as a couple and reduced the stress in the bedroom. I have heard some people say with the reduction in stress, their body changed and became more normal.

Soon after completing the state's requirements, a case counselor asked us to care for two brothers, Andy age six, and Drew age four. We quickly noticed the older one wanted to be with his parents, but the younger did not. This created much tension between them and escalated to extreme violence. Caleb, afraid for his life, would not sleep with his door unlocked. Using different things Andy said, I was able to ascertain he likely saw a relative, since jailed, murder someone. This and the horrible living conditions he suffered while at home probably led to his current behavior. We tried to convince the counselor to get him specialized help, but they refused because it would separate the brothers, possibly for a long time. I think the high cost was a factor as well. We were willing to adopt Drew. He would grab onto us when the caseworker would come take the boys on a visitation. Sometimes he would rip off his clothes so they could not go outside or wrap his arms around me to keep them from dragging him out the door. He always ended up going, but not willingly. One time the person transporting did not engage the child-lock on the car's back door, and Drew opened it while traveling in an attempt to not go on the visit. Thankfully, he did not fall out. I was not against considering adoption of Andy if they would pay for his mental health services. His young mind had a chance of being helped. We had no means to pay for his needed therapy and could not continue the violence in our home. I loved these brothers, and it broke my heart when I gave the ultimatum, either Andy gets help or he leaves my home. Unfortunately they placed them both with another couple, which also did not last. I fear they became lost in the system.

Then a counselor placed Enya with us, a seven-year-old girl. Abused at home, this sweet child had further abuse in other foster homes. She had behavior problems, but once she felt some stability and safety, she made substantial progress. The four of us grew to love each other. Instead of providing stability, the state reunified her with her mother. I would be fully on board with this if the parent was ready. She then bounced in and out of the system with no real stability.

After Enya, we had another set of exceptional brothers stay with us. They were in the process of being adopted by someone else. Then another boy, Arnold, whose mother regularly abused him, came to our family with the sweetest temperament. We adored him, and he loved us, but the state would not let him stay because they felt he needed someone of the same race. Family is not what color you are or who is related to you. Family are the people who love and care for each other and want involvement in each other's lives. The state thought we were good enough for him to stay nearly a year, and the counselor's, therapists, teachers, and his advocate all agreed he was thriving with us. I, yet again, was heartbroken to let him go.

While Arnold was with us, the state also placed Charles. In his mid-teens, he had all the normal hormone-induced problems of a teenager and years of emotional baggage from being bounced from home to home. Whenever he became close to a family, he would try to sabotage the relationship by acting out. This would eventually work and they would move him. This behavior directly results from knowing he could not stay with the family either way. With both parents in jail and no possibility of release within 40 years, the state refused to terminate parental rights so we could adopt him. Unbelievable! We all grew to love Charles, and I was able to manage his behavior for a while. He then started refusing to go to school and other places. He became more and more oppositional, and this caused one of us to be trapped in our home to care for him at all times. Eventually I think the case worker became tired of the constant calls for assistance, intervention, and adoption. I knew his behavior would improve if he knew he could have stability and family, instead they moved him to another home.

This was the last straw. We had been talking to other foster families in the county and found several who were only looking to help kids until reunified with their parents or adopted by others. They did not want to adopt, yet most of the kids placed with them were adoptable. Despite knowing we were hoping to adopt, the state seemingly placed kids with us which they had no intention of terminating parental rights while placing kids needing adoption with families not looking for adoption. If true, this is abusive and causes more kids to stay in the system. If this was not intentional, it is terrible mismanagement and a systemic failure on many levels. Either way, the kids were harmed by not having stability and it crushed my heart repeatedly.

Chance of Flooding

While Charles was with us, my business flourished with the big contract discussed previously. I opened an office in our small town's main street across the hall from an adoption consulting business. After they took Charles from us, Donna and I made an appointment to discuss the possibility of a private adoption with this company. We walked in for our appointment and my eyes widened into a big smile. The person greeting us was the wife of the executive pastor of the church we had recently begun attending!

After several discussions and prayer, Donna and I moved forward with a private adoption. We needed more home studies, background checks, and counseling. Then we submitted an exhaustive profile to several adoption agencies. The ladies, often still pregnant, wishing to place their baby up for adoption, work with an agency to review profiles and select those they think would be a good parent for their child. The agency contacts the selected person or couple to see if they are also interested. If everyone agrees, they formally make a match. At this time you have quite a bit of non-identifying information about each other. It took months before they matched one to us. Excited, I prayed for mother and child daily, and began to love this unborn child. Then the lady suddenly dropped the match and went with someone different. This is heart-breaking and much more so than I expected!

If things were not bad enough dealing with the adoption process, my office flooded along with the rest of the first floor of the building. Thankfully present when this occurred, I moved all the computing equipment out before the rising water ruined it with the help of a friend who operated a business upstairs. The furniture did not survive since it could not be moved immediately. I temporarily used a bedroom which we had used as a playroom once the foster kids left.

The agency matched us with another mother, who then later kept the child. Another match and failure because the lady ran off, was out of contact for a while, and then said she was going with a different agency. Then we had another match and a failure. Each one of these felt much like the first failure. However, each successive heartbreak compounded the pain of prior ones. We thought something was wrong with us each time they picked someone else after being picked. We still hurt from those past matches and this piled on top. All of this rejection and loss kept building and is in addition to all the foster kids yanked out of our lives after loving them so much. These losses were compounded by years of trying to have our own.

Questions surfaced if we should have had kids, any kids. We conceived our son through months of strong fertility medicine. We saw women blessed with pregnancy when we cannot, even though many were taking drugs, involved in prostitution, or otherwise had unwanted pregnancies. God will allow this person who cannot manage even the most basic of life's functions to have a child, but we cannot? Why? No really, both Donna and I yelled WHY to God! Many times.

> *"Lo, children are an heritage of the Lord: and the fruit of the womb is his reward. As arrows are in the hand of a mighty man; so are children of the youth. Happy is the man that hath his quiver full of them."*[1]

In Genesis 1 it says, "multiply, and replenish the earth". If

---

[1] *Psalms 127:3-5 KJV*

children are a heritage, a blessing, why are we being denied it? Why are people addicted to crack, meth, and other terrible drugs blessed with children, even multiple children? 1 Timothy 5 says we are to care for orphans, so why are we denied it?

These questions rumbled around in our head for years, but each month not pregnant the voice intensified. Each time the state would not allow us to adopt a child, it became louder. Now with each adoption match failure, the pounding scream soars higher!

Donna had enough! She was done. Over. Her confidence in humanity extinguished. Her faith in God crushed by the rocks on the shore of grief. She laid in bed crying often. I tried to keep her spirits up and keep her faith going, but I was struggling myself. I later learned my efforts were helpful, but I really felt like a colossal failure.

7

Madilyn

Faith is deliberate confidence in the character of God whose ways
you may not understand at the time. - Oswald Chambers

June 2008

I heard Donna's phone ring multiple times. A few minutes later, I heard it again. Curious about what was happening, that maybe she had walked away from it and did not hear it, I went downstairs. I found her in bed bawling.

"What's wrong?"

"Hannah's One Gift keeps calling. I do not want to talk to them. I cannot go through this again!" The phone rang again. "Don't answer it!"

A few minutes later I left the bedroom and almost immediately it was my phone ringing. It was the adoption agency again. Donna asked me to not answer her phone. I can answer mine, right? I went out to the driveway.

"Hello."

"This is Jean from Hannah's One Gift. I have been trying desperately to reach you guys. We have a child for..."

"Yes, we saw the calls. We cannot go through another match failure right now. We are emotionally spent. Donna is ready to abandon adoption entirely. We need..."

"You don't understand. We have a child. She is born and in our custody. Are you interested?"

"What? You have a girl, already born, AND in your custody?"

"Yes."

"I need to talk to Donna. Can I call you back?"

"Yes, but it needs to be very soon."

"Okay."

"Donna! Hannah's has a child already born. A healthy girl! They want to know if we are interested in adopting her."

"Really? She is born, and the birthmother wants to sign away her rights?"

"Yes! Jean said she already has guardianship."

"Oh, wow! Yes!"

"Hi, Jean? We are definitely interested!"

"Can you be here [over 1,000 miles away] in the morning?"

"Yes."

"I'll call you back when everything is confirmed for tomorrow morning."

"Awesome! Thanks!"

It is after 7 p.m. We would have to leave home at 4 a.m. to make a flight which arrived in time. I need tickets, but I cannot book them until I get confirmation and bags require packing for what typically takes two weeks or more. The kennels are all closed, so we must find other arrangements. So many things to do. A family going on a two-week vacation takes planning. This will be no vacation, and we cannot make a plan. It is time to do it!

At 10 o'clock, as concern grew, Jean finally called and confirmed. I booked the flight. We had already packed, and Donna found a friend to care for our dog. I crawled into bed, trying to rest for the busy day ahead.

We picked up our rental car and while driving to the store to buy a car seat and baby supplies, Caleb asked, "What race is she?" This question had not occurred to me. I would love her no matter what race she was, no matter how she looked. It made no difference. The only question I asked the agency, "Is she healthy?" The last time race was mentioned was during the questionnaire for our application to start this process. Neither of us had any hesitation except for laws allowing Native American tribes to take the child years after the adoption is finalized. I do not think Caleb cared either, but he was anxious about meeting his sister.

Then the adoption agency asked if we would be open to meeting Mia, the birthmother. They said she would feel better if she could meet us. I became very nervous. What if she did not feel better after the meeting? Maybe she would decide against letting us adopt her baby. On the other hand, refusing to meet might be worse, so we agreed to do it.

We showed up at the public park and waited. A little later an older woman walked up and introduced herself as Mia's mother. Mia was running late, so we talked with this lady. Eventually, Mia showed up. We talked about who we are and raising children. She wanted to know if we were open to correspondence and sending pictures. After answering her questions, she said she was happy we would parent her child and said goodbye.

We hoped the meeting went as well as she indicated but waited for the call from the agency. When they called about an hour later, they said all went well and they were arranging for the hand-over of our child.

Waiting around was excruciating and exciting. All day was a rush to something and then wait. The agency kept moving the handoff time later into the day. Then they delayed past dinner. After dark, we wondered if it was going to happen today. And then at bedtime we worried it would never happen. Wouldn't that be our luck? It certainly would fit our normal reality.

Almost 11 p.m. they called and said they were on the way to the hotel.

"Are you bringing our baby?"

"Yes, of course!"

Of course? In my mind, the question was valid after all which had occurred, but was immediately forgotten and replaced with elated excitement. The three of us quickly dressed and went downstairs to the lobby. Donna and Caleb sat, but I paced.

"Can I help you, sir?"

"We are adopting and they are bringing the child here in a few minutes."

It was a little more than a few minutes, but they walked in the door carrying the most beautiful baby girl ever born. I say girl because the most beautiful baby boy was Caleb.

My heart! Even now, twelve years later, my heart melts thinking of her. When Donna carried Caleb in the womb, I came to love my child more and more. When he was born, the love immediately magnified and overwhelmed me. Many parents spontaneously cry for joy at this moment.

Someone had warned me I would not love an adopted child like I love my "own". False! I did not get months to grow a love for my daughter. I had a few hours. Upon seeing her, all that love and all that occurs at childbirth, flooded into me, through me, over me. In the last twelve years, there has never been one moment when she was not my "own".

After Donna and I had taken a turn holding Madilyn a little, I heard a sound from across the lobby. The two clerks were sobbing with us. My questions on the timing and what took so long resurfaced. Something seemed wrong.

"What took so long to get her here?"

"We had to go to the state line to get her."

"What?"

I knew Madilyn was a few days old. What the agency failed to tell us, until this moment, was Mia had matched with another family through a different agency and had signed custody to them. Another couple had this girl, cuddling her in their arms, until a few hours ago. By law, a birthmother has a certain amount of time to reconsider and revoke the agreement. She invoked that right and then contacted our agency to place the child again. Sometimes women revoke the agreement because they want to keep the child. Way too often, they will revoke and then place a child for adoption again in order to get more money.

It is illegal to sell children, but it is common to ask those people adopting to help provide necessary health and living assistance. They asked us to help her earlier in the day. This practice is so routine it causes no concern about scams. Many people who feel like they must put a child up for adoption do so because they are struggling to support themselves.

Whatever her rationale for doing this, I was furious and heartbroken. This raised so many questions. How could the agency not mention this to us the day before? It felt deceitful. Would we have agreed to come? Was this woman going to do this to us also? How did the other adoptive family feel? I am sure they were devastated. Who would not be crushed by having their child taken after several days?

Of Course

We went to our room and tried to sleep. Well, Caleb and Madilyn slept. We spent the next few days being a family of four, discussing what to name her, and taking walks. Mostly, the time passed trying to not think about the what-ifs.

How do you not consider the possibility of someone revoking the adoption when it just occurred? Most adopting parents worry about this at least a little. Having found out my worst fear just happened to my girl, I am in a full panic. It must not be possible to hold a newborn too much, but if possible, I did it. I cuddled her as I prayed for her, prayed for us, and prayed for the family torn apart. I hurt for them and did not want it to happen to us.

The agency called, and I assumed it was regarding ICPC (Interstate Compact on the Placement of Children) paperwork. It was too soon for them to allow leaving the state. No, it was worse.

"Mia filed a petition to the court this morning to revoke the adoption agreement."

I was speechless.

"The judge denied the request. She submitted it after the time elapsed."

"So is it over? Does she have any other recourse?"

"It should be over. She did not file on time and the law is very clear. However, she is telling us she will sue and is getting a lawyer."

Of course she tried to revoke because that is what happens to us. After spending a week with this magnificent girl, I was going to fight to keep her. I even briefly considered jumping in the car and leaving the state. Since the police frown on kidnapping, I stayed.

More than a week later, both this state and our own completed the ICPC paperwork, so we drove home because flying with a newborn seemed much more difficult. If driving was easier, it was only slightly. She immediately fell asleep, but when time came to find a hotel, she woke up. Not wanting to spend the night awake in a room, not making progress nor getting sleep, I drove through the night to home.

We thought Mia had given up when the court rejected her. Soon after getting home, the agency told us she had appealed the previous decision and was suing for damages. Mia had a history of slipping in large retail box stores and suing them.

With at least forty-two prior filings, she knew how to get a free public attorney. Finding legal counsel who is fantastic in family court, well versed in adoption law, and experienced enough for me to believe they will win is daunting from a thousand miles away. Finding one who meets these criteria and is affordable does not occur. Thankfully, I found a lawyer who was not too rough on my wallet. Not only did he have a great background and experience, he had fought similar suits before, and he had adopted children. Perfect!

The judge expected us to attend multiple hearings spread over several weeks. The case was really about if Mia had filed the revocation appropriately and her complaints with the agencies. We agreed to be available via phone or video and fly out if a specific reason warranted it. Our lawyer successfully convinced him the financial hardship of flights and hotels, and flying back and forth with a newborn, was too much of a burden when we would not be needed. First hurdle cleared.

Ever since we had learned Madilyn's history, Donna and I researched adoption revocations. Donna found online groups discussing this with specific details, including one with Mia's name! I sent the details to our lawyer. With this information and her prior history of problems, we thought the judge would toss this case out. While he found it compelling and it helped sway the decision because it proved she knew how revocations worked, the case loomed over us for over five months. We could not rest until it was over for fear we would have to relinquish our daughter.

The trial ended in a victory for us, but overshadowed by the judge awarding Mia a small compensation. This meant I had to pay her attorney's fees in addition to mine, no small amount.

It took us several more months, when Madilyn's adoption finalization occurred, to relax.

8

Emily

Sometimes life hits you in the head with a
brick. Don't lose faith. - Steve Jobs

September 2009

"Thank You, God, for Madilyn's adoption finalization."
Like so many times before, standing in my driveway praying,
but this time it is all praise. "Thank You for my son, Caleb." I
had said many prayers before we became pregnant with Caleb
and said many more prayers for eight years after his birth
before coming to terms with not having any more biological
children. We then spent the last four years trying to adopt and
pleading for a child. "Thank you for my business, Donna's
new job, and our lovely home." Everything was falling into
place. For the first time in seventeen years of marriage, all
was perfect, but alarms should have been going off in my
head. Throughout my marriage, it seemed we did not get past

one trial before the next started. Yet all seemed perfect for over six months.

When the judge signed the adoption decree, many years of problems received closure. While life was not perfect, it was close. No other time could make this claim. Sure, more kids would be awesome. I think we both would like a larger family, but we were content. At 38 years old, this was the first 'perfect' time. We had two beautiful children who were healthy. I owned a company with contracts which paid well and loved every minute of work. After seventeen years of a marriage full of trials, we were a happy and content family.

My faith never wavered more than the common questions you would expect. I trusted in God, believed in prayer, and asked for a family. Maybe God wanted us to wait. Maybe He needed someone to help these kids by being foster parents. My ways are not like His ways. He uses us as He needs for His purposes and for our well-being. I still do not understand all which happened, but He was in control. I had two fantastic children and God provided all our needs.

A few months later in October, Donna nervously tossed a pregnancy test in my lap.

"You're pregnant? That's awesome!" I held back saying, "How?" After so many years trying, doctors had convinced us conception was not possible. I am stunned, then elated. Attempting to lighten the mood, "We can make this work but we will need a bigger house." Our house was more than large enough. This was incredible news. Donna said she was nervous because she did not know how I would react. Having two kids in diapers and waking all hours of the night would complicate things. I am the one who gets up with the babies so my sleep would be more strained for a while. I operate well on little sleep, and adding a third child now would be absolute perfection.

We still had Madilyn's baby furniture since she had only just outgrown it. The house was spacious and if we needed to juggle some rooms around we could. The home office could easily convert back into a bedroom. After the flood, I had moved my office into the spare bedroom, but continued to look for office space in our downtown area. I could give up the space at home. When the baby started sleeping more, the two of them could sleep in the same room until Caleb moved out so we did not need a bigger house. This birth would be simple to accommodate! Much simpler than an adoption placement of a baby already born! When we went to pick up Madilyn, we had nothing for an infant. We could not prepare because we were open to adopting an older child.

There is no way to overstate how elated I was. I prayed and pleaded before Caleb was born. We struggled for twelve years to have another child after Caleb. Having Madilyn's adoption completed was fantastic. Being pregnant was miraculous! Once the adoption placement occurred in mid-2008, I stopped praying for these things. I prayed over the children I had for safety, health, and their future. We became pregnant fourteen months after I stopped praying for another child. I pondered this and filed it away as an interesting side note of God doing things in His own time.

Even if the room would not be used as a bedroom for a long while, Madilyn needed a place to play when the baby slept. The flooded office had put a crimp on my business presence downtown, and dealing with the adoption problems left me too drained to work on a new place. After we adopted Madilyn, complacency had crept in while revelling in her. I started planning what I could do in addition to managing my contracts. One thing in foster care which came up repeatedly was a safe meeting location for parental visits. They often would end up at a fast-food playground. This is not ideal for spending time together. Also, I noticed many lawyers from out of town needing a place to work during the day between court appearances. I started looking for a place which I could transform into a few private meeting rooms and hopefully also an open co-working area.

After discussing with Donna how long to wait to tell family about the pregnancy, we decided to announce it at Thanksgiving dinner with my parents, brother, and sister-in-law. They lived 500 miles away and thus we did not see them often. Unable to do big birth announcements for the other two kids, I was excited to share this with them. Living so far away from any family, I always felt like an outsider. Infrequent phone calls and posts on social media delivered the occasional family news.

The occasion made me realize this was an opportunity to make a family memory! I wondered what scheme could surprise them with the news. It is nice to bring something to contribute to the meal; a decorated cake was considered. No, this would be hard to do well from so far away and impossible to hide once in the door. Then I thought about T-shirts with the four of us walking in with some obscure but telling depiction and see if anyone picked up on the message. When we visit them, we like to play Bananagrams, a Scrabble-like game making words out of letter tiles. I was settling on the idea of spelling out the message during an after dinner game. I'd lose the game, but it'd be a big win!

Friday before Thanksgiving, Donna had intense cramping. She went to the obstetrician and found out the pregnancy had ended in a missed miscarriage. Our baby girl, we later named Emily, had died! Donna had to wait until Tuesday morning to get a dilation and curettage (D & C). We did not go to Florida for Thanksgiving. Donna was still spotting from the D & C on Thanksgiving as my brother announced their pregnancy.

Emily's death had rocked me to my core. I broke. Many people have this idea she was just a miscarriage. Besides the child I so desperately wanted, Emily represented all we had struggled to have after twelve years of infertility. For God to give us Emily, show us our bodies were capable after seventeen years of trying for natural conception, and then have her ripped away, broke me.

We had the culmination of all our desires within our grasp. The pregnancy gave us hope! After all the problems we had while trying to conceive, this child made us feel like God saw us. The life in the womb represented so much of our life. We loved the girl, still love our daughter, but when she died our belief that God saw us also died. How could He tolerate this? Why would He give for which we had prayed and begged Him and then allow this? How could He be a loving God?

9
Year of Evil

Blind faith in your leaders, or in anything, will get you killed. - Bruce Springsteen

December 2009

I found an office for lease a little over 100 yards away from the county courthouse, with a private bathroom and multiple rooms suitable for private meetings and co-working. The private offices had windowed doors so parents could have privacy with their children yet allow the foster parents or caseworkers to observe interactions for safety. The rooms were large enough that they could hold small conference meetings for up to eight people. This also meant some kid's toys and games kept on hand would allow for an interactive area for parent visits. The space was exactly what I wanted. I signed the lease and started remodeling. Donna said I was burying myself in work to escape my grief. I had searched for the right place to carry out my big plans and the perfect office came open. For weeks, I spent every spare moment building out the space, painting, and decorating. This had nothing to do with Emily's death. Donna was right, I was only kidding myself. I over-worked myself again just like after Alex's death.

Five weeks after Emily died, the night after Christmas, a fire started three doors down from my office. The flames ripped through those businesses and burned out half my space. I had not finished renovations but a lot of furniture, equipment, servers, and other computers were already in place. I had started using some rooms for my software consulting.

Donna saw posts on social media and then the news. She told me and I ran into town. It was the morning after Christmas and should be quiet. There was a mob surrounding the office building as firemen looked through rubble.

Unbelievable.

After running up to the fire command trailer and getting permission to go in, I rescued what I could, including most of my important documents. Much of the equipment was burnt and warped, but not my primary work desktop and monitors. I picked up the chassis and about a gallon of water spilled out. I took it anyway, hoping the hard drives were still operational. Very little else was salvageable. A wall was missing, and what remained unscorched the firefighters soaked in chemical-infused fire-fighting water.

I went home, opened the computers up in the garage and laid all the papers out on towels to dry inside the house. Once this chore was completed, I said to Donna, "We need to leave town."

I was exhausted emotionally. We packed our bags and spent a few days at an indoor water-park a few hours from home. I could not wrap my head around all that happened in the last few weeks. We had ten months of relaxation after seventeen years of problems. The miscarriage, D & C, and a major fire within five weeks. Our knack for attracting trouble has kicked into overdrive.

Roller Coaster Fireball

February 2010

The D&C procedure failed to prevent an infection. Additionally, Donna's endometriosis became much worse. She had severe cramping and ovarian pain. The infection caused constant bleeding. She had to wear hygiene pads constantly. If someone bumped into her, she immediately needed to excuse herself. Physical intimacy was rarely possible. Her doctor said she needed a hysterectomy.

Unfortunately, the insurance company disagreed. They thought a uterine ablation could be sufficient to cure the problem. I found their belief disturbing since they never did an examination, but doing the surgery without insurance was cost prohibitive. Because of the life-threatening health risks associated with getting pregnant after an ablation, and because we just proved a pregnancy is possible, the doctor required me to get a vasectomy as a precondition to her doing the procedure.

So in a few short weeks, we obtain hope and excitement from a surprise pregnancy. Then our child died, lost a business by fire, and then find out the miscarriage exacerbated things to the point neither of us would ever conceive again.

> If our hopes went on a roller coaster ride, then the coaster holding our hopes derailed and crashed in flames. If Emily's death had rocked me to my core, this took the core and split it.

I had so many questions. How could He allow this to occur? If He allowed it, then He was involved. Either He actively took Emily or actively did not stop it. If He is all-knowing, all-seeing, and all-powerful, then He knew this would result and could have changed the outcome. He could have prevented the pregnancy in the first place. He could do it or He is not who He says He is.

This rambling around my head had to stop. Not because of where it led or my sanity, it must cease because Donna needed me present. She had so many of the same questions and I desired her present. We had a tough road to walk. Her body needed to heal, and we had not even begun the journey.

Who?

Let me pause the story for a moment. Have you noticed

anything about this story which seems a bit odd? Is it the long list of trials we have endured? No, our trials may be difficult, but I know others who have similar stories, and those who have endured much more.

This story has yet to reference one person coming alongside to help since 1991, nineteen years prior. There was no family nearby with which we could commiserate. No one was around to babysit Caleb if we needed to have a moment together, or even to watch him as we went to a doctor's appointment. We went years between date nights. Even when we were doing foster care, at least fifty weeks a year we were in church. We would ask friends and people we trusted at church to babysit with pay but had no one to accept. We led a small group for our church. I played in the orchestra at church and later ran the audio, while Donna taught Sunday School. I coached U8 and U10 soccer, and was Cubmaster for two years of the Cub Scout troop and active in scouting for years. We were actively volunteering and helping others, but had no social life and no help.

I do not know how we survived foster care, and adoption legal battles, and running a business full time throughout while Donna taught school. The government refused help when we needed food. The government refused to facilitate our adoption. Family lived hundreds of miles away. All of this, and more, which is not in this book, without simple help.

We were going through these things alone, and it felt like isolation. No one was nearby. We contacted our church, and they were not interested in helping or praying with us. Our small group quit attending suddenly, everyone at once. One of them said they were not allowed to meet with us. Who disallowed our meetings, and why? A leader called Donna and asked her to teach Sunday mornings again because teaching the five-year-olds would help her get her mind off the child she lost. Being immersed in a room of little ones will not distract one from losing a child. This might not be true later, but especially not the case in the early days of grieving. He did not even ask how she was doing. She was still recovering from the previous medical procedure, and even a random stranger with this information might wish someone well. This lack of concern or prayer was a shock to me.

How can people who come into our home weekly, suddenly abandon without a trace? We studied the Bible together and went on day trips into the mountains together. Every person I served beside weekly for years in church and scouting vanished. Not having the support of those who worshipped God with me demoralized me more than the others.

Donna was not recovering, emotionally or physically. I kept trying to prop her faith up and ignored my crisis of faith because she needed to have some hope. Our bodies heal better with hope and I knew if she completely lost faith, hope would follow. Donna's doctor, a well-regarded specialist, did the ablation. Just as she predicted, the procedure did not work. Months went by with Donna struggling to do simple tasks through the abdominal pain. We both were battling through the emotional pain. Eight months after the failed D & C, the insurance company approves Donna to have a hysterectomy. They prolonged our suffering for far too long.

Someone at church finally decided that all these health problems were real, and maybe they should bring a few dinners after Donna had the hysterectomy. After eight months of needing prayer and friendship, we get six meals, one every other night.

I get angry every time I think about it, not because they brought dinners, but it was not the thing we truly needed. I am a skilled cook and could provide meals, but appreciated the food as it left more time for me to help raise Caleb and Madilyn. We wanted someone we could lean on; someone who would listen. At the very least, one who would entertain Madilyn while I straightened the house or had business calls.

Decade of Evil

August 2010

We hoped the hysterectomy would get our lives back to some semblance of normal. It relieved Donna's abdominal pain, and with the infected organ removed, her general health improved. However, this ordeal left her with a problem causing recurring genital and rectal skin fissures. Any movement would irritate the affected area. Walking was painful. Routine cleaning , simple bowel movements, and gentle intimacy would painfully aggravate the issue or cause new tears. Her doctor and other specialists sought to find the solution. Donna tried many medicines and procedures to no avail.

I can not imagine living with this pain daily, and I saw it. You can not stop basic body functions. Everyone requires cleaning. She could not stay immobilized to reduce irritation for very long. The one thing possible to control was shutdown; physical intimacy was yet again nonexistent.

All of those questions rambling around in my head screamed for attention.

NOTE: Finally, five years after it began, a solution was discovered to stop the tearing in Donna's groin. In 2015, the doctor found a way to repair the issue. It required an out-patient procedure at the hospital. As I sat waiting for the surgery to finish, I got a call from my mother. My grandmother had died. I hated that I could not attend her funeral. At least Donna would finally get relief from this torture.

Well, this procedure worked and kept Donna out of pain for five years before resuming once again in 2020. The reminders of the Year of Evil ten years ago do not stop.

10

What Faith

*Faith and doubt both are needed – not as
antagonists, but working side by side to take us
around the unknown curve. – Lillian Smith*

Near the end of 2010, I started questioning God's character.
Did God take all our desires, serve them up on a silver platter,
then throw them onto the ground so He could stomp on
them? What purpose is there in tricking a couple who spent so
many years trying to get pregnant? Why make them think
conception is possible, then allow the child to die? When we
were content to have the two children God gave us, why stir
the pot and raise our hopes? Why then take those desires and
destroy the body such that no possibility will remain?

*"Ask, and it shall be given you; seek, and ye shall find;
knock, and it shall be opened unto you: For every one*

that asketh receiveth; and he that seeketh findeth; and to him that knocketh it shall be opened. Or what man is there of you, whom if his son ask bread, will he give him a stone? Or if he ask a fish, will he give him a serpent? If ye then, being evil, know how to give good gifts unto your children, how much more shall your Father which is in heaven give good things to them that ask him?[2]"

"For every one that asketh receiveth." Did we not plead for years? I understand getting an answer of "No." or "Wait." But when the desire of over 4,000 days is gifted, I did not expect the stone and the serpent! This verse made me wonder. I think I am a good father. If my child asked for something, and I said, "No", but then later allowed, "Ok, you can have it.", would I smack it out of their hands? Would I then hurt them for wanting it?

In many places in the Bible, it refers to us as adopted heirs. We are His children. In other verses, the Bible calls us the bride of Christ. Both adoption and marriage have strong legal ties in the Bible and speak to how God demands we treat our children and our betrothed. In Matthew 11, shouldn't anyone feel confident replacing the names so the Bible verse says to them personally, "if his son, David, asks for bread, will God give him a stone?"

A parent has every right to say "No" when it is appropriate, but when is it ever proper to give a stone when bread is wanted. I felt like God gave me a big, ugly stone. He placed the

---

[2] *Matthew 11:7-11 KJV*

desire in my heart to ask for children. When I am finally offered hope with a living child in the womb, she is replaced with a hysterectomy instead.

Is God love? Does He love others, but not me? Am I not really a Christian? Maybe I think I am, but am fooled.

How can a God who loves me do this? He could have just left it as a "No". I could have gone on my way with my two children, continued volunteering at church and community activities, and witnessing to others about His goodness.

Maybe God was not who He claimed He was. Perhaps God is not who I think He is.

"Who are you?!"

When Moses asked, God said "I Am that I Am" in Exodus 3:14. I heard nothing, though.

"WHO ARE YOU?!"

I begged this repeatedly. Not as a fool yelling into the void, but as a man wanting a response. I was not only standing and waiting for a voice, yet I did that some. Searching for an answer, I studied scripture for a year. I was no stranger to reading His Word, but I focused on studying.

"Do you exist? Are you out there somewhere but not here?"

All the while, I am helping Donna through her very painful groin issues. I have tried to stay active in scouts and Caleb was nearly an Eagle. Madilyn was learning ballet, and we switched to another church. I was volunteering running the audio and Donna was teaching Sunday School again.

Life looked normal, at least to outsiders. My contracts are going great. Donna is working in a new position at the local high school and taking post-graduate studies to get her next advanced degree after Master's.

Inside the home, and inside our heads, the questions never ceased.

What Happened?

September 2012

From the traffic light, I turned onto the highway and continued half a mile. The fall leaves had begun falling, but plenty were still in the trees. They blocked my view around the turn. A car had stopped in my lane and my motorcycle laid down, sliding underneath the vehicle. Okay, this recreated scene based on the police report lacks many details. I recall sitting at the stoplight, a few seconds of clarity while being wheeled into surgery the next day, then laying in the hospital two days after the accident wondering what happened. My helmeted head had hit the road hard and knocked me out, unconscious, in and out for thirty-six hours. The wrist shattered as it went through the rear-view mirror. Sliding tore my ACL and a litany of minor injuries. Thank God I had my business and worked from home. The many injuries constrained me to the bed and couch for months while recovering.

Then, within one year of the accident, I lost all of my small contracts. Quickly thereafter, the big contract I had for seven years dropped all further work. No one seemed to want my services and none were hiring, or at least not hiring me. I even had several refuse to hire me because I was overqualified. We had big bills due and savings were low. The adoption, lawsuit,

Donna's ongoing medical issues, my physical therapy from the wreck and associated expenses, and our willingness to give anytime God told us to give, all contributed to a lean account. Months went by without me having a paycheck.

I finally found a job as an employee with a massive pay cut. I had to sell my boat to pay bills, and eventually we had to sell our home and move to a smaller house. It took over six years to pay off all the debt incurred. We called 2010 the Year of Evil and called 2012-2013 the Year of Evil II.

After I resumed my search for truth, I reminded myself of Job questioning God. Then I got mad. Not for the first time, this was just another reason. Job went through a lot. Job lost seven children, all of his wealth, his wife, and had rough health. No joke, if his story is true, it was all kinds of messed up. However, he came back around, and his life restored. Along the way he had friends staying with him, supporting him, trying to help. He became wealthy again and grew old, surrounded by many children and grandchildren. My wealth was gone, our health was bad, and I did not get to have a large family. Donna did not have any support other than me. I had one friend who would talk while fishing once a month. Later, another friend would call about weekly, chatting mostly about finding faith. Everyone else, including our local faith community and family, was absent.

Reset

During our weekly phone calls, my friend and I would talk

about various issues around how the Church had moved away from the teachings of Christ and the example set by the early church as recorded in the book of Acts. The conversations then turned toward Deconstruction, a term adopted by many who pull apart their traditional faith looking for answers. After a few months of exploring, it became apparent this group has most of their ideas grounded in relativism. Some may argue they ground their truth in scripture, but I saw none whose foundation did not invoke a cultural ideal of goodness or used an unmerited historical context. Relativism, by definition, would not lead to an absolute truth. No amount of introspection or debate will convince me of a truth that originated without evidentiary support. Also, it seemed like many in the movement were trying to validate their preconceived ideas of what they want to be true. Many espousing their thoughts on it started with an idea they thought true and looked for a way to rebuild their faith around it. Many adherents considered their doctrine too rigid. Some thought Hell could not exist, or there was not enough love, or the Bible contradicted science, or that God would not hate sexual immorality. I had questions on some topics, but I wanted truth no matter which way it led me. I had no destination in mind other than the truth.

Mike McHargue, well-known thought leader amongst many Deconstructionists, has axioms about faith. This is one of many which I tried to consider, but ultimately dropped because of the subjectiveness.

> *"God is AT LEAST the natural forces that created and sustain the Universe as experienced via a psychosocial*

> *model in human brains that naturally emerges from*
> *innate biases. EVEN IF that is a comprehensive*
> *definition for God, the pursuit of this personal,*
> *subjective experience can provide meaning, peace, and*
> *empathy for others."*

I cannot operate with this axiom. Either God exists or does not. I will not worship a set of natural forces, even if it created all that exists. If an entity formed everything, then I would want to know if it is a personal being or if I should consider this some experiment set in motion. If the Creator is not personal, not interacting in some way, I won't worship it. Why should I? However, if there is a Creator and if He is personal, I do not want to be against Him.

How can I keep searching for God and questioning if He exists? I thought I had no faith left, but this is not true. I must have had faith to continue to think He exists, or might exist. How can I prove this to myself? I am a mathematician with a logical mind. Belief in God might never have a rational proof but maybe I can find a a credible reason which is not in contradiction to hard facts.

> *"Now faith is the substance of things hoped for, the*
> *evidence of things not seen.*[3]*"*

Whatever I discover, I would like to frame it mathematically. If faith is the substance of hope, then maybe it is not concrete. Court trials deal with this all the time and call it a "preponderance of evidence". Mathematically, it is proving

---

[3] *Hebrews 11:1 KJV*

there is a better than 50% probability of a thing being true. Where do I start? The Bible says God exists. How can I prove the Bible is true? Literary analysis and prophetic studies try to answer this, but those are not the sciences I understand well.

> *"So God created man in his own image, in the image of God created he him; male and female created he them.[4]"*

Evolutionists say man emerged from other life forms, which evolved from single-celled organisms. I knew evolution was too difficult a place to start looking, but how did life begin? How does a cell form? How were the complex chemicals created? I studied this for months and decided I needed to go back further.

> *"In the beginning God created the heaven and the earth.[5]"*

Science says in the beginning there was a Big Bang which formed matter and all which we now have. I supposed the beginning was a good place to start.

---

[4] *Genesis 1:27 KJV*
[5] *Genesis 1:1 KJV*

11

Faith Foundations

*Science means constantly walking a tightrope
between blind faith and curiosity; between
expertise and creativity; between bias and
openness; between experience and epiphany;
between ambition and passion; and between
arrogance and conviction – in short, between
an old today and a new tomorrow. – Heinrich
Rohrer*

You need to have a little background information on our past, the way back past to the beginning. I am sorry this needs to get technical, but will try to keep it brief. The basics of how we understand our universe formed is fundamental to relating it to how we got here. If you will stay with me, hopefully you will find interesting beauty and design.

Many theories have existed explaining how the universe

came to be. About one hundred years ago, a paper proposed the theory of an 'explosion' which created the universe. This later morphed into the Big Bang theory. While this idea based its assumptions on observed movement of the galaxies and laws of physics, no solid proof existed until Penzias and Wilson discovered the cosmic microwave background radiation, an electromagnetic energy remnant of the Big Bang.

Many scientists think our existence came accidentally through chaos. They accepted the Big Bang as a massive burst of energy and matter, flung across the universe, and through some chance, formed hydrogen and a bit of helium atoms. After a few hundred thousand years, the atoms settled enough to be stable and clump. Some of these clumps of atoms became massive enough so their gravitational attraction caused them to form a ball and squeeze into the center. If the ball was large enough, the pressure would raise the temperature sufficiently so the hydrogen in the center would start burning. It is now a star. This burning is really fusion, and it is what we see from our Sun.

As a star consumes the hydrogen, it fuses it into helium. Once the hydrogen is nearly gone, the star could begin fusing the helium if there is enough of it present. The ones who do are called red giants. Not all stars are large enough and become white dwarfs. However, few red giants are the right size to become a supernova. We need supernovas though to create the much heavier elements needed for planet creation.

Normal star fusion creates elements like oxygen, carbon, and iron. When a star explodes and forms a supernova, the pressure and temperatures fuse the elements into gold, uranium, lead, tin, iodine, zinc, and many more. The explosion throws these elements, and a lot of simpler ones, through space.

Our solar system was likely formed by a cloud of these elements collapsing to form our sun. This kicked material such that it formed asteroids, moons, and planets. Our sun is likely made from the remains from a second or third generation supernova. Have you seen pictures of the Earth next to the Sun? It looks smaller than this period by the capital O. This is how our sun would look next to the massive stars which formed our sun.

Also, let us cover a little seventh grade mathematics. The probability of tossing a coin and it landing heads up is one out of two, or 0.5. This is half the time. In this book I will discuss some tiny probabilities which will be hard to express as a decimal so I will use scientific notation. For example, if you have ten dollars in pennies and mark one, then ask someone to pick it out of a jar while blindfolded, their probability of selecting the marked penny is one in one thousand. This is $1/1000$, which is 1×10^{-3}. The exponent is the number of zeros, and the negative sign means it is a fraction. If you see 1×10^{-5}, then think $1/100000$, one chance in 100,000.

The magnitude of the exponent is a tricky thing to understand. Each single digit increment is ten times bigger than the previous. The difference between 1×10^2 days and 1×10^3 is that the former is just over 3 months and the latter is almost 3 years. Let's add one more to the exponent, 1×10^4 days is over 27 years! The numbers get big very fast.

More Foundations

I find it odd one would think the random chaos from the biggest explosion which has ever occurred in the history of ever, would produce anything as complex as a planet. Why does this seem odd? Let's start with the laws of thermodynamics, highly simplified:

1. No energy can be created, but it can be transformed.
2. All energy exchanges reduce the potential for work. Hotter things warm cooler things and everything disperses until temperature and dispersal are in equilibrium.
3. As things disperse, they cool down to a state of not moving.

These are fundamental laws governing the physics of the universe. Unlike speed limits, they cannot be broken. Starting with the first law, if no energy can be created, then where did it come from to make the Big Bang? The world's greatest minds have been working on this for many years. The renowned expert, Stephen Hawking, came up with an answer[6]. He said our universe could have gotten its start from another universe or dimension. It is the conclusion of the Multiverse Theory. If this is true, then from where did the energy in that other realm come? Hawking was also the one

---

[6] *Hawking, S.W., Hertog, T. A smooth exit from eternal inflation?. J. High Energ. Phys. 2018, 147 (2018). https://doi.org/10.1007/ JHEP04(2018)147*

who previously thought[7] the universe had no beginning nor an end. It would expand and contract, renewing itself repeatedly. This is similar to the cyclic model, which even Einstein considered for a while. Both theories run into problems with the second law of thermodynamics, and for this reason most physicists do not consider it now. I presume this might be why Hawking moved towards the Multiverse.

My elementary analysis already has one rational explanation favoring some sort of creator. I would not place this as definitive, but it raises large doubts on the prevailing scientific explanations.

Looking at the second law, it would seem that the initial burst out of the energy release would have dispersed, cooled, and continued to do so. It would have if it were not for gravity, and electro-magnetic force, and properties of particle attraction, and many other major details.

I knew, way back in my brain, the laws of thermodynamics and how the universe needed some kind of push to get it going. With so much focus on my pain, I failed to step back and realize I already knew a Creator must exist and at the same time was angry that He allowed it. The fog of grief had clouded my ability to think or see the roots of faith which remained. Faith so tenuous, it needed more proof, and it must have convincing evidence the being is personal and involved. Even so, I only allowed myself to say for the purposes of

---

[7] *Hartle, J.B., Hawking, S.W. Wave Function of the Universe. American Physical Society 1983, 12 (1983). https://link.aps.org/doi/10.1103/ PhysRevD.28.2960*

proof, "A Creator must exist and is whatever force which created the universe." This is fundamentally not even faith as a deist; even the multiverse fits this definition. How do I prove more?

Scientists talk about the Big Bang, but there are four such generally accepted monumental leaps.

- Creation of energy and matter in our universe ("the" Big Bang)
- Creation of life-sustaining planets, like Earth.
- Creation of life.
- Creation of sentience (humans) able to question if they were created.

Insert Magic Here

Regarding the Creation of Energy and Matter, I have addressed the other prominent scientific thoughts which either kick the can to another realm or describe how the universe has always existed.

The idea of a multiple universe theory and that some other universe acts upon ours is an extraordinary claim. For me to buy into it will take exceptional evidence. I have read through the ideas and what I am left with is Hawking and others trying to explain why our universe is so fundamentally geared towards order and life when the laws of thermodynamics are against it. He is looking for a way to escape the need for a creator but has no explanation past conjecture rooted in quantum field theories. There are other arguments, like we are inside the black hole of another universe and our black holes might lead to other universes, but these have no justification either.

The theories stating the universe has always been here, like the cyclic model or Hawking's shuttlecock supposition, suffer from violating the first and second law of thermodynamics. How did the energy and matter come to be in the first place? How does a universe expanding out at an enormous rate like ours, come to violate the second law and implode and gain back its potential energy?

I cannot find a well-thought theory which can come close to any beginning state which does not involve 'insert magic here'. Whatever this magic is, I am back to "A Creator must exist and is whatever force which created the universe."

Probably

The creation of life-sustaining planets was briefly discussed earlier. Gases form stars, stars die and some go supernova, and heavier elements form then eventually collapse into a new star, and hopefully, a solar system is born with planets. Given a lot of time, this seems plausible. However, the details make it very unlikely to occur, and when it does, it is inordinately uncommon to support life.

Remember, the second law of thermodynamics makes it such that all things want to disperse, break apart, and cool down. The energy from creation flung material away from the origin. We see this momentum even now. Most galaxies are moving away from ours and away from each other. The reason things did not immediately fly off on their own, the reason the first stars were created, relates to various other forces at work.

Our universe exists on the edge of a knife. Any little change, one slight change to any of many special numbers, could end life immediately. Even though I thoroughly disagree with Stephen Hawking's conclusions leading to infinite universes, he was a brilliant scientist who admits our universe is special.

> *"The laws of science, as we know them at present, contain many fundamental numbers, like the size of the electric charge of the electron and the ratio of the masses of the proton and the electron. ... The remarkable fact is that the values of these numbers seem to have been very finely adjusted to make possible the development of life."*[8]

One such number is the cosmological constant. All matter has a gravitational pull. This force keeps your feet on the ground and makes the leaves fall. Our moon orbits the Earth and does not fly off because the Earth's gravity pulls it. When a rocket takes off, the engines are providing thrust greater than the force of gravity so the rocket moves away from the Earth. There is a force negating the effects of gravity in our universe. It keeps the universe from collapsing, all matter coming together into what would be similar to a black hole. If the force were any stronger, stars and planets would never form. Instead of collapsing, the universe is expanding but not enough to stop planetary formation.

---

[8] *Stephen Hawking, 1988. A Brief History of Time,Bantam Books, ISBN 0-553-05340-X, pp. 7, 125.*

"*Cancellation to that accuracy is impossible to comprehend. If we take the total number of sand grains in all the beaches of the world and charge half of them with positive charge and the other half with negative charge so that charge cancels out, producing a neutral pair, this cancellation is only to some 24 decimal places... Even if we consider the estimated total number of elementary particles in the observable universe, we would get 'only' 80 decimal places. A cancellation to 120 decimal places requires extremely delicate fine-tuning!*"[9]

The force is ever so slightly more than the gravitational pull, and this allows the universe to expand. The forces must equal for the first 120 decimal places. Imagine a tug of war and you measured the pounds per square inch each side exerted in their pull. The losing side would only have to pull a little harder to tie; the weight of one atom! This tug of war had equal forces to only 25 places.

Other constants needing extremely precise values are the electromagnetic force, strong nuclear force, and gravitational forces. These three have nothing to do with each other, none influence the character of another, yet work together to keep a perfect balance. If gravity were much weaker, stars and planets would not form. If it were only slightly weaker or electromagnetism only slightly stronger, the stars formed

---

[9] Calle, C. I. (2009). The Universe: Order Without Design. *Amherst, N.Y.: Prometheus Books. pp. 156.*

would be cooler and not explode into supernovae[10]. Without these, we would not have the heavy elements allowing life. The strong nuclear force strength relative to electromagnetism also must be balanced. If the difference changes by half, few elements other than hydrogen could form, and likely no other but hydrogen. The universe would be a boring place! If you think a 50% change is a lot, there is less leeway than it sounds; even extremely small differences from where it is would have dramatic consequences. Scientists did not know how stars could form so much carbon from stellar fusion but must since it is abundant on Earth. Recently, a physicist proved it occurs only with the current balance in these forces within a certain temperature range. A small deviation between these two forces would keep us from having large quantities of carbon or oxygen[11].

All life on earth is carbon based. Plants, trees, moss, algae, fish, whales, elephants, flies, humans and every other living thing requires carbon and oxygen. Plants produce oxygen (O_2) but they need carbon dioxide (CO_2) for photosynthesis, which needs oxygen. Hypothetically, life could be based on silicon or other materials instead, but most scientists agree those could not withstand advanced life and are problematic.

---

[10] *Carr, Bernard J. and Martin J. Rees, 1979, "The anthropic principle and the structure of the physical world", Nature, 278: 605–612. doi: 10.1038/278605a0*

[11] *Hoyle, Fred, D.N.F. Dunbar, W.A. Wenzel, and W. Whaling, 1953, "A state in C^{12} predicted from astrophysical evidence", Physical Review, 92: 1095.*

Oberhummer, H., A. Csótó, and H. Schlattl, 2000, "Stellar production rates of carbon and its abundance in the universe", Science, 289(5476): 88–90. doi:10.1126/science.289.5476.88

These constants of physics in our universe each have a relative probability of it being within the range allowing life. When I was studying this in 2011-2013, I did not have all the resources which are available now. Many physicists had published likely probabilities, but finding layman-accessible articles explaining them was difficult. Maybe the problem was I did not know what I was supposed to be finding. Of what I found, I did my own rough calculations of the compound event probability and came to a very conservative 1×10^{-160} probability against the universe creating planetary systems.

Recently, I found a paper by Luke Barnes[12] using Bayesian theory testing which finds the probabilities of the Plank constant and cosmological constant being within the range of even supporting structure in the universe. The combined probability is 1×10^{-123}. Barne's numbers support my early calculation being incredibly conservative and that the combined critical constants' probability is likely much smaller. Additionally, in 2008, Hugh Ross did a detailed analysis[13] of items required for life once you assume the universe's physics work the way they do. He calculated; galaxy cluster formation of the correct type, galaxy placement and form, planetary system placement and form, and planet composition. He also removed dependencies and reduced the value based on known assurances. His compounded

---

[12] *Barnes, Luke A., 2012, "The fine-tuning of the universe for intelligent life",* Publications of the Astronomical Society of Australia, *29(4): 529–564. doi:10.1071/AS12015*

[13] *Ross, H., 2008, "Why the Universe Is the Way It Is". :Baker Books. Appendix C, Part 3. Probability Estimates for the Features Required by Various Life Forms*

probability is 1×10^{-811}. I am willing to assume Hugh missed some dependency or some other error occurred and the probability is not this small. If the Barnes' and the modified Ross' probabilities are compounded, a highly conservative 1×10^{-800} odds would insanely improbable.[14]

Incomprehensible

How small of a chance is this? Let's play a popular lottery with the numbers 4, 8, 15, 16, 42 and 23. Assume you gambled once a month with the same numbers each time for forty-two months. How many of those do you expect to win? Winning every single one of the forty-two lotteries[15] is better odds than the universe not collapsing and produces fundamental elements heavier than carbon, thus life might occur. Do you think this is doable? Consider just the combined probability of the two constants Barnes looked at. If you formed one-hundred-trillion dimensions and made one-hundred-trillion machines in each dimension, which could each make one-hundred-trillion universes every nanosecond. If these machines tried no combination twice, it would likely take well over one-thousand-trillion-trillion-trillion-trillion-

[14] *Ross' probability exponent is -811 and Barnes' is -123. This makes their combined probability total 1×10^{-923}.*

[15] *$(3.305 \times 10^{-9}$ odds of one lottery win$)^{42} = 1 \times 10^{357}$ which is less than $(1 \times 10^{123} * 1 \times 10^{311})$ {Barnes' calculation & Ross' odds for simple life}*

trillion[16] years to find one universe[17] which did not immediately collapse or expanded without making stars. Again, this is only considering two of the constants. We cannot wrap our head around this number, much less how improbable considering the other constants.

If you continued playing the lottery every month for ninety-four months (seven years and ten months), winning every single lottery, you still have better odds than having a planet capable of sustaining advanced life.

There are more critical constants which must align so the universe exists. One might understand these values being so precise with each other if there were some intrinsic relationship between them. If there is, no one has found it, and many are looking intensely for any correlation.

Once the basic complexity of the cosmos' existence hit me, and having known some obstacles of cellular creation, faith reached the tipping point of absolute belief in a Creator. Here I realized there must be a Creator who had the intellect, creativity, and power to balance our universe on this knife's edge. The Creator with this much power could create without the need for extreme precision. The only rational conclusion is if the Creator did this for a reason, it was to show people like me He exists.

I then became intrigued by how far this design went. Did God really make life, or was it a random spark which evolved?

---

[16] 1×10^{63} *years*

[17] *Actually still a one in three billion odds against finding one... (1×10^{14} dimensions) * (1×10^{14} machines) * (1×10^{14} universes) * ((1×10^{63} years) * 365.2422 * 24*60*60 *(1×10^{9})) is less than 1×10^{123}.*

Did He breathe into Adam or was he a figurehead for the evil which men had become? Had God sent His son to be the Messiah, or was he just a good man? How much of what I had been taught did God leave proof through empirical evidence?

This is where it really hits me. I did not lose my faith at all. My trials had knocked out all the old foundations upon which I built my belief. I had doubts, and I still had many questions.

12
Faith Creation

Faith means belief in something concerning which doubt is theoretically possible. – William James

The hard part of creation must be behind us, right? We have a stable universe, stars, and a mixture of elements exist. There is plenty of energy with low entropy to make planetary systems, and at least one of these systems created one planet capable of sustaining life. No, it turns out that even though the universe is extraordinarily rare and Earth's existence should be improbable, creating life is much more difficult.

The interesting thing about having questions of faith, I was not afraid to question anything. If God is True, as in existing and as in being the standard of truth, then why be afraid of the answer? Before I started studying the creation of the universe, I wondered how life came to be on this planet. Many

Darwinians believe with enough time and the right primordial soup on enough planets, life must occur. I read all I could find online and came across a video of Stephen Meyer. He was talking about his book, Signature In the Cell[18]. This book made me realize I needed mathematical assurances God exists.

Meyer almost lost me; I nearly put the book down because the beginning is heavy on philosophy and history. He writes well, but philosophy is not a favorite of mine. Then the book moved into science and mathematics.

Water is necessary for all life, and in the absence of water, life could not occur. Most evolutionists believe life, the first cell(s), started in a primordial soup; a sea of water rich in fundamental building blocks. One cell is a complex arrangement of millions of proteins working together with lipids, carbohydrates, and glycans.

The first problem to overcome when creating life in water is how to form peptide bonds to link the amino acids together, which then form the protein. Peptide bonds are extremely unlikely to form because water[19] attacks the bond. To solve this, there must be specialized enzymes to make the bond occur. The paradox is, how do you have enough enzymes to make proteins when enzymes are a type of protein?

Once a protein happens to form, the problem then is what protein will it make? Even with all these things in the primordial soup and assuming we can get past the water

---

[18] *Meyer, S. C. "Signature In the Cell"*, 2008. pp 204-216.
[19] *Williams, A., & Hartnett, J. W.* (2005). Dismantling the Big Bang: God's Universe Rediscovered. *Green Forest, AR: Master Books.* pg 159.

issue, random amino acids are not likely to bond in an arrangement which is useful.

Most proteins have many hundreds or thousands of amino acids joined together. Meyer likens it to a chain of those colorful childhood linking blocks where each block is an amino acid and the knob joining blocks represents the peptide bond. Imagine having twenty different color blocks to choose from. Let us try a small chain of only 150 amino acids. This is 20^{150} ($1x10^{195}$) combinations but only 1 in $1x10^{74}$ are a valid functional choice according to Meyer and others. When you are dealing with exponents, 74 is not even close to half of 195. It is a tiny fraction of a fraction. While there are many proteins, many are not functional, and of the ones which are, only a fraction are useful.

Even if there was a combination of amino acids, which happened to occur in the correct order to create a useful protein, it is unlikely to get the correct bond. In the lab, amino acids form peptide bonds only about one-half of the time. Also, the amino acid bonds come in identical mirror images, a left-handed version and a right-handed version. I am left-handed, but I can use right-handed scissors, while a functional protein cannot. They can only use the correct-handed version.

The odds of each pair being left or right-handed is roughly half. This makes the chance of all being correct $(1/2)^{149}$ or about $1x10^{-45}$. The possibility of peptide bonds occurring are about the same. Ignoring the overwhelming number of amino acids which bond and do not form a useful protein, the decent candidates only have a $1x10^{-164}$ odds against finding just one foldable protein. This is yet another staggeringly massive number. The lottery is getting boring and the odds are too easy in comparison! Assume I had marked one grain of sand and you had one chance to pick it from all the sand in the world, from the sand dunes of the Sahara to all the beaches, without looking. Think how impossible this is. Now imagine you picking it correctly eight times in a row! This is easier than finding a single functioning protein.

The film "Origin", by Illustra Media[20], describes the difficulty of finding a single foldable protein. Imagine the Earth's oceans filled with the twenty different amino acids and having the perfect environment for making a protein. With no contaminates or other problems, we also make it so the proteins quickly self-assemble. This perfect environment allows six-thousand-million-billion-trillion-trillion protein chains to form every minute. The longest estimated age of the Earth is 4.6 billion years, and in that time 1.10^{58} chains would form and likely disintegrate immediately because they cannot fold. This is so far removed from the chances needed to likely find a foldable protein that if you had one amoeba traveling one foot per year, it could cross the observable universe and back before one foldable protein would be created. Additionally, this amoeba could then transport one atom at a time across the universe and finish moving every atom in the observable universe to the other side before one foldable protein was made. Oh wait, it can then move the entire universe fifty-six-million more times! See their website

---

[20] *"Origin", directed by Lad Allen, narrated by Alvin Chea, Illustra Media, 2016*

(http://www.originthefilm.com) for more math details and clips explaining it much better.

I have read great explanations of how nature would try to overcome this via natural selection or memory. The math does not add up. In Meyer's book, he discusses Bill Dembski's work on maximal opportunities to find an event since the origin of the universe. He postulates that using the known observable universe as the maximum area of effect is a large enough sized region since the speed of light would keep any other particle from affecting what occurs here. The generally accepted number of elementary particles in the observable universe is approximately 10^{80}. The fastest these particles can interact is 10^{43} times per second, and since the beginning of time has been about 10^{17} seconds, there could be at most 10^{140} total events in the observable universe. This is a much smaller amount of time than what is needed to find one protein!

The odds are so overwhelming that when asked is it unlikely that species arose the way Darwin said, or is it impossible, David Berlinski [21] responds:

> *"There is hardly a difference. We are talking about odds that are so prohibitive. If you wish to say 'impossible', fine, I'll defend you saying it is impossible. If you wish to say it is 'highly unlikely', I'll be in your corner as defense attorney as well. There is no practical*

---

[21] *Performance by Peter Robinson, et al.,* Mathematical Challenges to Darwin's Theory of Evolution, *Hoover Institution, 22 July 2019, www.youtube.com/watch?v=noj4phMT9OE.*

difference."

When an organism replicates, simple DNA alterations may not affect the organism, but mutations needed to create new species need coordinated changes in the DNA and in the discrete Gene Regulatory Networks (dGRNs). A dGRN is a collection of molecular regulators which interact with other GRN in a hierarchy, and with other substances in the cell. They govern the gene expression levels of mRNA and proteins and thus determine the function of the cell. Any attempt at altering a dGRN kills the organism. They must be built from the ground up, but this requires new proteins.

David Gelernter wrote a review called Giving Up Darwin[22] on Stephen Meyer's book, Darwin's Doubt[23]. In the review he states:

> *"Evidently there are a total of no examples in the literature of mutations that affect early development and the body plan as a whole and are not fatal."*

What you need to see is, for intelligent design to not be true, proteins must be built from random processes in extremely harsh conditions, not the ideal conditions in our example with the amoeba above. Not just one protein, but one cell has millions of protein strands and thousands of types. This massive set must be within a few microns of each other and the proper lipids and carbohydrates nearby in order to come

[22] *Gelernter, David. "Giving Up Darwin." Claremont Review of Books, 2019, claremontreviewofbooks.com/giving-up-darwin/.*

[23] *Meyer, S. C. (2013).* Darwin's Doubt: The Explosive Origin of Animal Life And the Case for Intelligent Design.*New York: HarperOne.*

close to being a cell. The cell must have DNA and RNA, which tend to have millions of base pairs. Human cells have billions of base pairs. The simplest cells have over 2,000 proteins and at least 182 DNA base pairs. If the materials for a cell did occur and were thrown together, make sure they were all created within a short time so the RNA does not decompose. In normal Earth temperature ranges, RNA lasts only a short while so be sure the proteins which aid in replicating the RNA are around, and the proteins which help with cellular division have been discovered also.

Some will try to trivialize these probabilities by claiming there are many other planets trying for life, and this just happens to be one which won the lottery. First, we see above the numbers are so huge against life, that even if each atom in the universe could try for life many times a second, they would need trillions of trillions of years to find the first two proteins. These proteins may not work together, and the first one likely decayed a trillion years earlier, so it is back to square one. How do you get millions of correct proteins together? This completely ignores the need for a valid RNA strand which is even more difficult to build correctly.

Then if it hits the jackpot and makes a cell, and the cell replicates, how does it have the ability or remaining time to modify genes and dGRN? If it makes a new species gene strand, it has to be done millions of more times. Critics of Intelligent Design say the problem is solved by massive parallel organisms all trying all the time. Remember, we have already incorporated every particle in the known universe, it cannot be more massively parallel. Even if there were multiverses and many had stable universes, the odds of a single cell ever forming randomly are mind-boggling unlikely. Yet, here on Earth, after the crust cooled down to a reasonable life-sustaining temperature, the first fossilized life was deposited in a few million years. A few million years is not much time at all!

The Darwinian theories are based on hundreds of million of years spent on cellular mutations using processes we have never seen, never replicated, with no intermediate proof, and have no good basis to know how DNA base pairs can be added while altering the dGRN and the organism survive. I am mostly ignoring these Darwinian theories because so far we have all but proven it is near impossible to even build the first cell. Without the first cell, these theories are meaningless. The Earth went from too hot for life to complex life in 30-50 million years[24]. This is way too improbable for me to grasp. As Berlinski said, it is no practical difference from impossible.

Atheists Convince Me

Having briefly shown a small bit of the complexity of even having a universe exist, this book then showed how creating the first proto-cell has overwhelmingly tiny odds of success. Between the Big Bang and the first biological life are the enormous odds of our planet existing in a life-sustaining area of the universe which was glossed over in the last chapter. After biological material somehow existed and happened to be located in an area smaller than a grain of sand, it overcame

---

[24] *Staff (20 August 2018). "A timescale for the origin and evolution of all of life on Earth". Phys.org. Accessed 20 January 2021.*

gargantuan odds when became a working unit together and replicated itself!

> *"You can construct an argument; something short of a proof. I agree if the [roulette] ball drops in the slot marked seven just once, and you get a big payoff, 'hmm, that's interesting, dropped once'. If it drops in the number seven again, and again, and again, and again, at some point you would be entitled to scratch your head, and say, 'You know, the game must be fixed, or I must be inordinately favored to be winning it like this.' But I'm not going to say, 'I'm lucky.'" [Berlinski]* [25]

I believe I have sufficiently proved a preponderance of evidence that a Creator exists, at least I did to myself. He is all powerful and has effected a marvelous creation. Its beauty transcends from the numbers on these pages, to a walk on the beach, and gazing at the stars. Each view shows He is the Creator.

If life did not start by random chance on this orb, then He

---

[25] *"David Berlinski - Atheism and Its Scientific Pretensions." Performance by David Berlinski, and Peter Robinson, YouTube, Hoover Institute, 1 Sept. 2011, www.youtube.com/watch?v=FyxUwaqooRc.*

created sentience as well. The Bible's book of Genesis was written over 5,000 years before we knew anything about how the universe was created, yet it gives a very accurate representation.

This was not the end of my faith journey, far from it. I started this in 2010 during our Year of Evil. It took me three years to hit my rock bottom. I was still questioning WHY! But when I saw God's creation and the undeniable proof it is His, I took pause. I had to step back.

"Who am I to question what You do? You created the ground I am standing on, the rock hurtling through space, and the Sun which keeps me warm. The Bible calls you King and Lord, rightly so. If nations can bow to some man and call them king, if they can do his bidding without question, even so I can do yours."

Simple servitude is all I had. I could not wrap my mind around love. We went through so much and several years after, Donna was still fighting her medical issues. I had lost my contracts, and several business ventures collapsed. We were barely surviving.

It took another two years of intense searching scriptures and studying literary analysis for me to trust my interpretation of the Bible again. Only then could I find a renewed love of God. These two years were full of various studies, but you can get a lot of it just by watching all the videos you can find by Gary Habermas and by finding Dr. Bart Ehrman's views on the historicity of Jesus. Dr. Ehrman is fascinating. Even though he is an atheist, he proves Jesus was real, and that the people who saw him believed he performed miracles.

The best evidence we have proving Jesus lived, according to Bart Ehrman, is fifteen independent sources for the crucifixion of Jesus within one hundred years of the event. He says he knows of no accredited specialist in the field of religion and actually teaches in universities, colleges, or seminaries, who doubt Jesus lived.

The second best evidence we have is the Apostle Paul. Accredited textual critics and historians generally agree[26] that seven of Paul's New Testament books are authentic. They can verify authorship and historical accuracy. In these we get the story of Paul, in Galatians 1 and 2, meeting with James, the brother of Jesus, Peter, and John the Beloved. They meet to discuss the nature of the Gospel to make sure they are preaching the same message. This makes Paul an eyewitness to the eyewitnesses.

Virtually no accredited scholar denies Jesus lived because they have proof he died. If he lived, then his followers either believed in the resurrection or foolishly spread the story, knowing they would most certainly be killed for the message. They killed John the Baptist before Jesus. Shortly after the crucifixion, Stephen and James, brother of John, were martyred. Threats kept coming at those preaching the Gospel, and eventually all the apostles died torturous deaths except John the Beloved. I do not believe someone who did not have extreme evidence would stand in the face of certain death instead of being silent.

---

[26] *Ehrman, Bart (December 16, 2014). "Pauline Forgeries: 2 Thessalonians as a Test Case". ehrmanblog.org/pauline-forgeries-2-thessalonians-as-a-test-case/ Accessed 20 January 2021.*

Paul was one of the people killing Christians when he had the vision of Jesus! It was two years after the resurrection and the story was strong and the early creeds already formed. Paul did not go to meet with Peter, James, and John to create the Gospel. It was already formed and Paul had previously been trying to squash it. James, the brother of Jesus, was not a believer until after the resurrection and he saw the risen Christ. Why would he convert suddenly while knowing it would make him a target also unless he thought it was real?

Those who say the apostles faked the resurrection have big hurdles promoting the theory. People around Jerusalem would all know of the crucifixion. It would be common knowledge where they placed the body and it would have had guards posted, just as scriptures say. If the disciples faked it, they had to sneak by the guards and somehow steal the body. They knew the rulers would kill them for saying he had risen. This makes no sense at all unless they believed He had risen. There were also over 500 people who thought they saw him after death, and some felt his wounds. How do you fake this? Otherwise, the leaders trying to stop this movement would have pointed to the dead body and proclaim he is not alive. This did not happen.

This was my last hurdle. God must be a loving God to have gone through this much trouble. I had no idea where this left me, though. I did not feel loved.

13

Caleb

Sometimes you find your strongest faith in the darkest corners. - Margo Price

2018

First, an elder arrived and our friend soon after. In total, five couples from church showed up quickly after we made contact. Having these people be present gave us so much strength! Even though they may have thought they did not know what to do, just doing what they could was perfect. None of them had knowledge of what we were experiencing, nor did they know what to say sometimes. They prayed with us, talked with us some, cried and was a shoulder to cry on, and sat in silence together. I cannot imagine surviving this day without them present.

They heard what we wanted regarding a service, made some calls, and then told us what was available nearby. With our approval, they made all the arrangements for the funeral and arranged for the paperwork to be brought to the house for us. They made the arrangements for using our church and set up the space.

Over the first few weeks, various ones invited our daughter out on play dates, to movies, anything to let her get away for a moment. It helped her to get out and relax emotionally. Her play dates allowed us to not worry as much about her mental health.

A stark difference occurred in the response of losing Caleb versus the response we had upon losing Emily. Some friends showed up this time. It was not the nature of the death which caused the different response. I have seen these people, and others in our faith community, rally around all types of loss. This is just what they do. This is love, care, and compassion.

If you have not experienced this gut-punch wailing from the loss of your only son, and then walked the lonely desert of grief, then you cannot imagine[27] how much everyday life assaults me. If you did not lose your child to suicide, you do not suffer the accusative stares of those who think parents should have known it would occur. I have my own feelings of guilt to deal with. I do not need more thrust upon me. It is not just stares; some ask "How did you not know?" and "Were there signs?" Even the passive questions are assuming a parent must know. Even the experts say it is impossible to

---

[27] *I acknowledge other losses can hurt deeply and no two losses can be directly compared. Losing a child to a disease without rationality and coupled with social stigma has more torment than I can express.*

always know.

It has been over 1,000 days since he died, and I feel no reduction in grief since the first day. The only thing which has changed is I can control my reactions better, at least those outside of my body. I do not sob uncontrollably very often, but I still do. A few weeks ago I was crying during church service because something the pastor said reminded me of the enormity of my loss.

I cannot remember where my glasses are most of the time. At least once a day I prepare coffee or water and I leave it in place, forgotten. Going to the store almost always causes anxiety or PTSD symptoms. The most basic life tasks are arduous.

After we lost Emily, with nearly everyone abandoning us and everything in my life falling to pieces, I still wanted people. However, I could not stand seeing people avoiding us. For several years I hated being forced into isolation, now it is my refuge. Being around people now causes anxiety and stress. I became an introvert.

I think a lot of the isolation became ingrained within Caleb. He was thirteen when Emily died and he hated how no one came to our house to socialize. He even begged people to come over. It was not long before he started being an introvert as well. I hate this because it may have contributed to his problems dealing with life. He felt alone, but his independence kept him from moving home when things became difficult. I look at Madi who is now almost the same age and wonder if she will have the same issues.

I have questioned God a lot since Caleb died.

"WHY????"

"Why must I endure yet another loss? Each loss is exponentially worse than the last! Why did You allow Caleb to die?"

I have poured my heart out, brokenness is all which remains. I had Emily ripped away. Even though I still mourn for her and Alex, I had made peace with the situation. I was making a life with my wife and two kids. Now, yet another child is taken.

The one thing which has not happened this time is questioning who God is or His sovereignty. This question was settled years ago. I know people who lose a child to suicide and have to face these questions of faith, wondering who is this God. I am glad I do not have to process both things simultaneously.

To you who are questioning Him, I offer you hope. God is waiting patiently. He can handle your tough questions and is not afraid of your doubt. He works in the doubt. If there were solid answers with definitive proof, it would be called math, not faith.

God exists.

God is personal and caring. It would take me another book to explain how and why, but this is yet another area I think the modern church gets wrong. He is not a kumbaya, touchy-feely, Jesus. He has the unconditional love of a perfect father. Jesus' parable of the Prodigal Son shows this relationship well. The child wants to leave home and see the world. After warning of the dangers, the father gives the child a full inheritance, all he would receive if he had stayed until the father died. When the child had wasted the money on partying, he lived in filth. Eventually, he went back home, not being able to survive on his own. The father, culturally shamed for raising a foolish child and expected to disown him, having every right to be upset and turn the child away, ran to his son. He embraced him and gave him a place to live.

Lastly, even though those around you bail, God is always here. Everyone close in my life bailed when we lost Emily. I made new friendships. Nine years later, when Caleb died, we lost many more close relationships, but not all, and we found more. There are good people and even when they are not near, God is.

I know it does not seem like He cares at times. It does not always seem as though he is near. I know this feeling well. I also know it is not good to trust our emotions, especially during the difficult days, even years, after a traumatic loss.

People care and if you do not know any who care, I do. I know what it is like to feel this way and you can reach out to me.

14
Faith in Action

In the absence of any other proof, the thumb alone would convince me of God's existence. –
Isaac Newton

Possibly the biggest lie in existence, the one most often told, is "I'm fine." I am guilty of saying it. Using fine in this sentence means to be satisfactory or pleased. It may come from the French word "fin", to be complete or whole. Neither of these meanings describes me most of the time.

So why do I say it and do not mean it? I will revert to this familiar greeting when I am tired of explaining my mental state or do not know the person well enough to explain it. Yes, our culture has degraded this lovely expression of care to a simple greeting that rarely wants a real response. Many times I will respond with "Hello."

"But I would not have you to be ignorant, brethren, concerning them which are asleep, that ye sorrow not, even as others which have no hope."[28]

"And we know that all things work together for good to them that love God, to them who are the called according to his purpose."[29]

I resist saying something more truthful because of the responses I often receive. These verses, and more, have been used as a club. People are not trying to be mean. They want to help and use them hoping others will feel better, however, it often has the opposite effect. Additionally, these verses are being taken out of context.

In Thessalonians, Paul encourages us to not be sorrowful like those with no hope. This does not mean we are not to have sorrow. We are sad our loved one is no longer with us, but also have hope they were redeemed and we will see them again.

When someone quotes Romans 8:28 in an effort to comfort someone who lost a child three weeks earlier, it comes across as cruel. The first thing I thought was I would rather have my child than whatever purpose God wanted. Then I wondered if God took my child to achieve some purpose. Both roads lead to immensely dark places. Seriously dark; I struggled under the weight of it for months. I do not believe God took my child, yet He could use the situation for good. My experience has helped several people choose to keep trying, to not give

---

[28] *1 Thess 4:13 KJV*
[29] *Romans 8:28 KJV*

up on life. If this is all the good which occurs, it is enough.

Not Just Another Monday

I have always regarded myself as a Creationist, though I shied away from the label. Traditionally, it meant one who believes in a Creator. Now, people infer you are one who believes in a young earth creation; a literal six sequential calendar days to create all of the cosmos and humans as well. I believe in the literal creation story, but not in one calendar week. I believe the Bible is inerrant and divinely inspired, and I interpret it literally where it should. The story of creation is in a section of historical literature, and thus should be read literally. I believe in a Creator who has created all things and did so in six days. However, I do not believe those days are sequential, Earth-based, calendar days.

This is a key area Christians have become lazy with their understanding of scripture. Genesis tells a beautiful story of creation, relationship, rebellion, and redemption. However, we either gloss over the first point or get mired in arguing it.

I discussed the tip of the iceberg of creation earlier in this book. My knowledge is crude at best, yet it is much more than most people know because few people teach it. I believe discussing scientific creation will lead many to Christ, who ordinarily would not consider it.

The opposite is also true. Dismissing the scientific view of creation puts many people off of Christianity. I would never argue for a theory based on popularity. On the contrary, I argue for scientific methods because it is based on evidence observed in our world. If God created the world, then the evidence points to Him.

Many Christians read Genesis 1, and then believe it shows how creation occurred in one, seven day, calendar week. This naïve reading completely ignores what the scripture is trying to convey and ignores fundamental basics of our universe, physics, biology, and mankind.

It is not just modern Christian scientists, like Hugh Ross[30], John Lennox[31], and Stephen Meyer[32], who believe in a day-age theory of creation. John Wesley, Saint Augustine[33], Aquinas[34], Origen, Irenaeus, and many others have proposed theories of creation along the same line of the day-age theory. The day-age theory states that God created all things but took some period of time to do each task, not necessarily a calendar day. Irenaeus was a well-regarded church leader in the second century. He studied under Polycarp, who was a disciple of the Apostle John. Irenaeus stated creation was a long period of time. Aquinas said, "not from a want of power on God's part, as requiring time in which to work, but that due order might be observed in the instituting of the world."

These early church fathers prove the idea of an older Earth is not a reaction to recent scientific findings, nor is it a way of

---

[30] Ross, H., *"A Matter of Days : resolving a creation controversy"*. *Reasons to Believe*, 2015.

[31] Lennox, J., *"Seven Days That Divide the World"*, Zondervan, 2011.

[32] *"The Age of The Earth, by Dr. Stephen Meyer"* Performance by Stepen Meyer, YouTube, Colson Center, 30 April, 2010, https://www.youtube.com/watch?v=GGFWH6Okgl8

[33] Augustine, "The Literal Meaning of Genesis".

[34] Aquinas, T., "Summa Theologica".

shoving evolution into Christian thinking. Many young earth believers say any other belief is denying the Bible, not taking it literally, or call it heresy.

The scientific evidence for an old universe and an old Earth is overwhelming. God does not lie or deceive. He is Truth. I would need extraordinary proof He went out of His way to deceive us into thinking the universe is billions of years older than it is. We see galaxies which are hundreds of millions of light years away. This means it would take that long for the light to reach us. We can prove the earth was formless and void like Genesis says when it was a molten blob and later as a water world.

> *"The heavens declare the glory of God; and the firmament sheweth his handywork.*[35]*"*

> *"For his invisible attributes, namely, his eternal power and divine nature, have been clearly perceived, ever since the creation of the world, in the things that have been made. So they are without excuse.*[36]*"*

He would not hide what He had created; the heavens declare it! His creation is clearly perceived, so no one has an excuse when they stand in judgment. Why would He make the world in one week, make it look like it took billions of years, then say it was fast in Genesis, and then say in Romans He would use it to hold people accountable when they had not been told

---

[35] *Psalms 19:1 KJV*
[36] Romans 1:20 KJV

who He is? I do not think a God who is True would do this.

Reading the Bible in English is fraught with error. This is a fundamental problem with translating Hebrew and Greek to English. We can lose some nuances. I am not saying the Bible is wrong. It is us mere mortals who sometimes do not understand what is written and sometimes make errors translating what they wrote. Unfortunately, few people do the appropriate work needed to know what the Bible really says. When we study what the scripture means, we still can get it wrong. We are only human.

One way we get things wrong is by assuming the meaning of common words and phrases. We are used to reading modern writings; novels, news, textbooks, etc. The Bible was written thousands of years ago in an unfamiliar language. Even if you grew up speaking Greek, it has changed over time. The first Bible translation in English was by William Tyndale in the early 1500's. As an example of how quickly languages change, consider Tyndale's first 5 verses of Genesis, translated into English 500 years ago.

> *"In the begynnynge God created heaven and erth. The erth was voyde and emptie ad darcknesse was vpon the depe and the spirite of god moved vpon the water. Than God sayd: let there be lyghte and there was lyghte. And God sawe the lyghte that it was good: and devyded the lyghte from the darcknesse and called the lyghte daye and the darcknesse nyghte: and so of the evenynge and mornynge was made the fyrst daye."*

Only 100 years later, we have the King James version. The lettering changes, the spelling is wildly different, and the sentence structure changes.

> *"In the beginning God created the heaven and the earth. And the earth was without form, and void; and darkness was upon the face of the deep. And the Spirit of God moved upon the face of the waters. And God said, Let there be light: and there was light. And God saw the light, that it was good: and God divided the light from the darkness. And God called the light Day, and the darkness he called Night. And the evening and the morning were the first day."*

It is astonishing how quickly the language changed. Languages change often, and not just how words are spelled, but what those words mean. In the King James' version, it says, "fear the Lord" many times. We now translate the word fear as one of respect, honor, or revere. Really, it should be all of these words at once, but our language does not have a good parallel. However, the change in translation is not because it was wrong when translated in 1611. The connotation of the word has changed over time.

Let's revisit the assumptions made when reading. In the original language, Hebrew, it does not say exactly as the King James version does.

> *"... the evening and the morning were the first day. - KJV"*

> *"... and there is an evening, and there is a morning --*
> *day one. - YLT"*

This is significant because while this pattern occurs on days one through five, it then when creating humans, it changes to
"...-- day the sixth"

Then the next verse, "he rested on the seventh day from all his work which he had made." In Hebrew, one counts things with the article 'the'. The first five are written like a parenthetical. The Hebrew of the Bible had little punctuation or even vowels. It does have the definitive article, like the word 'the', but no indefinite article. So to indicate an indefinite amount of something, it tends to not use an article. Since days one through five do not use the definitive article, they are likely meant to be longer periods of time. John Lennox[37] proposes the first five days are real actual days but separated by some time to allow God's command on each day to come into its own. Creating the universe could look like a Big Bang to us and then nine billion years later, God spoke light upon the Earth on the first day. The first verse creates the universe, but it is not a part of day one.

So, yes, I believe quite literally that God created all which is, including man. I also believe He used time and processes. These processes led Darwin, with his limited knowledge, to think adaptations led to the origin of species. I do not believe this interpretation of creation leads to Godless Evolution. I think it shows very clearly how God directed and created all things without needing to answer exactly how it was

---

[37] *Lennox, J., "Seven Days That Divide the World: The Beginning According to Genesis and Science". Zondervan, 2011.*

achieved.

Doubt

Did I lose my faith? For a while, it felt like I did. I questioned everything. It was not just my faith that I questioned. Grief will do that. Loneliness will do that, and I was alone. I felt abandoned by God and most people I knew. I had an occasional talk with a friend, but, at best, I was drifting aimlessly for way too long.

As I tumbled down the rabbit hole of questions, there were no answers to cushion my hits on the rocks and roots along the way. I had accusations flung at me constantly. Some accusations came from those in church and some from myself. The condemnation from both generally revolved around my lack of faith, and how I should blindly ignore what occurred in my life. I should smile, be happy, and act as if nothing was wrong.

We were told if we had faith, if we prayed as we should, God would bless us with another child to replace the one we lost. I stared at the person as Donna acted like she was holding on hard to something inside. This was a few months after Donna's hysterectomy. This person knew all about it. I have no idea what kind of miracle she thought was going to occur.

I had a zealous, completely free of doubt, faith for most of my life. Even back then, I would be hard-pressed to think God would perform the miracle of birth without a uterus. He could, no doubt, but would He? Unlikely. This person was not being prophetic or claiming this was from God. It was a passing remark as she welcomed us into the church. It is a statement made from being told the same things I was told for years. She probably did not mean to upset us with the remark, but it hit us hard.

Remember the end of chapter two? "Blindly trusting in what you believe is like being on the water with no life jacket. Everything is comfortable when it all goes smoothly, but when life throws something at you, you may want something to hold on to tight." We had blow after blow thrown at us. Without the foundation of a proper understanding of our faith and the stress of needing to do everything alone, we eventually lost our footing.

In retrospect, why would I have thought I had a good footing in my faith? I thought I did. I went to church regularly, prayed, and studied my Bible. Any time someone asked me to serve others, I was there. However, I did not have a firm foundation in faith even though I thought I did. If you think you are good, that your faith is sure, how do you know? Have you asked yourself hard questions and forced yourself to answer them? I had the book knowledge and the answers the Church continually pushed, but it is not the same as living it through trials. I had faith before, and I knew God created everything, but during intense grief, the questions flood in.

I have heard pastors say, "I know because I know." How do you know? "I know because the Bible says it." Do they think this will convince someone else? Doubtful. When asked for a reason, the response should be grounded on something firm, either experienced or logical. The Apostle Paul had an encounter with the risen Christ on the road to Damascus. Some people are rooted in the many fulfilled prophecies. I had to come through Mathematics. I am not saying these people are wrong in their emotions, but if this is their sole answer, they have no real foundation.

> *"But sanctify the Lord God in your hearts: and be ready*
> *always to give an answer to every man that asketh you*
> *a reason of the hope that is in you with meekness and*
> *fear"*[38]

The word 'reason' here is 'apologia' in the original Greek. It is the application of logic to persuade someone. This is the same as giving a defense in a court case. In English, we have many other definitions for reason which water down the idea that we are to convince others with rational, yet polite, argument.

When I have encouraged others to take a serious look into knowing why they believe, they have told me this is not Biblical. Instead, I am admonished that "Faith comes by hearing and hearing by the Word of God."[39] Faith is the evidence of things hoped for. Yes, the Word is crucial, but faith needs a foundation to believe the Word is true. And so, here is where reason comes in. When we are telling others about our faith, telling the Gospel, and giving a reason for our faith, others gain faith in His Word. If the Bible alone was enough to bring faith for everyone, the Apostle Peter would not have said we should always be ready to give an answer. Peter would instead have said to be ready to read the Bible.

> *"For the invisible things of him from the creation of the*
> *world are clearly seen, being understood by the things*

---

[38] *1 Peter 3:15 KJV*
[39] *Romans 10:17 KJV*

that are made, even his eternal power and Godhead; so that they are without excuse."[40]

How are the people without excuse if they must hear the Word of God to believe? How are they to know of Him unless our belief can be obtained from creation? The Holy Spirit works within us and opens our eyes to the design and the Designer.

The Church has become lazy. It takes study and work to be ready to answer people's questions. The Church needs to push teaching Christian apologetics as a primary discipleship discipline. The lackadaisical attitude toward faith and apologetics has kept many from accepting Christ. Unfortunately, it also drives Christians away as well.

"Many believers are afraid to ask for help because they've been told depression reveals a 'lack of faith.' So they suffer silently. Then, if they slip over the edge and kill themselves, everyone is surprised.[41]"

Do you know who else suffered from a lack of faith? Biblical faith heroes Abraham, David, Peter, Gideon, Elijah, and even John the Baptist. They had help to overcome their doubts. We

---

[40] *Romans 1:20 KJV*
[41] *Grady, Jay L., "Do People Who Commit Suicide Go to Heaven?". Charisma Magazine. (2018) charismamag.com/blogs/fire-in-my-bones/38437-do-people-who-commit-suicide-go-to-heaven, last accessed February 2021.*

need to do better at helping others overcome doubts and lift them up.

The Apostle Thomas doubted the reality of the resurrected Lord (Luke 24:38 ; John 20:27). The Greek word for doubt here, apistos, can mean one who never had faith or one who is questioning faith. Thomas just spent the last few years living with Jesus. His doubt was not a denial or unbelief, but an attitude or feeling of uncertainty. Jesus does not severely rebuke Thomas, but neither is his skepticism commended. "Stop doubting and believe" is the word of the Lord to his disciple, but He said in love. Earlier in the story, the Apostle Peter declared his undying devotion to the Lord, and then a few hours later had lost faith and denied knowing Jesus.

There are different meanings of the Greek and Hebrew words which are translated into English as 'doubt'. Like the word love, it has many connotations which need context and study to know what the Bible is telling us. I believe that doubt (disbelief) keeps us from God, but doubt (questioning) can bring us closer to God.

Doubt is not the same as disbelief. We often feel guilty for allowing doubts to pop up in our minds, especially if it's about something we thought we had a strong faith in. But doubt is not disbelief, and you shouldn't feel guilty about it. Disbelief is a mindset of, "This is not true," while doubt is a question- "What if the things I believe aren't true?" Of course, doubt can lead to disbelief, but it can also lead to a stronger faith. It all depends on how you react when those thoughts of doubt come up.

Doubt can strengthen your foundations of faith. Doubt, honest questions put forth looking for truth with an open mind, often puts your faith through a test. If you put in the work to find the truth, then truth will come and you will come out stronger for it.

Pursuit

Christians do not judge the Apostle Peter harshly for his stumble. They do not regard Abraham or John the Baptist as unbelievers. Abraham became the father of the Jewish nation and the Bible lists him in Hebrews 11, "Faith Hall of Fame". The Catholic Church considers the Apostle Peter to be the first bishop of Rome, and thus, the first pope.

> *"Be it known to you, my lord, that Simon [Peter], who, for the sake of the true faith, and the most sure foundation of his doctrine, was set apart to be the foundation of the Church, and for this end was by Jesus himself, with his truthful mouth, named Peter."[42]*

Would Abraham had continued and had faith after being incessantly berated for having slept with his servant because he doubted his wife could become pregnant? Would Peter preach the Gospel if people continually reminded him how he denied Christ three times? Yet, when someone like myself has questions and looks for evidence, we are bullied or treated poorly.

> *"When I speak to college students, I challenge them to find a single argument against God in the older*

---

[42] Clement, *"Letter of Clement to James 2"*, A.D. 221

agnostics (like Bertrand Russell, Voltaire and David Hume) or the newer ones (like Richard Dawkins, Christopher Hitchens and Sam Harris) that is not already included in books like Psalms, Job, Habakkuk and Lamentations. I have respect for a God who not only gives us the freedom to reject him, but also includes the arguments we can use in the Bible. God seems rather doubt-tolerant, actually.[43]"

God is more than "doubt-tolerant". He seems to go out of His way to comfort those hurting in their doubt. Most people hurting and doubting are questioning, "What if He does not exist?" God continuously pursues us, showing us who He is, so we might be with Him.

In 1 Kings 19, the prophet Elijah ran for his life, sat down under a tree. Depressed, he asked God if he could die. Then, like most of us when depressed, he fell asleep. Later, an angel came, woke Elijah up, and told him to eat food that had been prepared. Again, like us, Elijah ate the food and went back to sleep. The angel came a second time with more food, woke Elijah, and had him eat.

When life gets tough, sometimes we need to sleep and have meals prepared for us. The greater the trauma, the more we need help. Elijah had been doing God's work and saw many miracles, yet he despaired and was ready to die.

Before Christ was to be crucified, He went to the Garden of Gethsemane to pray.

---

[43] *Yancey, Philip., personal website, philipyancey.com/q-and-a-topics/faith-and-doubt, last accessed 6 March 2021.*

"... and began to be sore amazed, and to be very heavy; And saith unto them, My soul is exceeding sorrowful unto death... [44]*And there appeared an angel unto him from heaven, strengthening him.*[45]*"*

Jesus' grief was so hard to bear, God sent an angel to help. Why are the grieving and hurting abandoned as unable to be helped? When Donna and I needed help in 2009-2012, those closest to us and who we served, abandoned us. Maybe it was too difficult to help us. However, it is incumbent upon Christians to be self-sacrificial; serve when and where it is needed. Having a solid support group would not have removed the grief but, as 2018 showed, it smooths the road and makes the arduous journey easier. The Church should be like ministering angels when people need it, and often it is! When it is not, the damage can be terrible, even irreparable.

Why are the grieving and hurting given poor advice? I believe this comes down to a poor understanding of scripture and this is from laziness. Laziness hurts yourself and those around you. A Church community relies upon each other to build people up.

I was lazy as well. Before tragedy struck, I should have done the hard work of knowing why I believed what I believed. I had studied mathematics, biology, chemistry, and astronomy. With this knowledge, I could have dug a little deeper and put more of the creation story together sooner. I ended up needing physics and cosmology for it to become clear, but I

---

[44] *Mark 14:33-34 KJV*
[45] *Luke 22:43 KJV*

doubt I would have spiraled down as I did had I put in the work. I also do not think I would have doubted as much if I had support.

The combination of being raised, always afraid of losing my salvation, coupled with the pervasive reminders of doubt being the opposite of faith, caused me to think my doubts meant I was not a Christian. I wondered how I could doubt so much and be saved. But then I would remind myself of the Bible verse,

> *"For I am persuaded, that neither death, nor life, nor angels, nor principalities, nor powers, nor things present, nor things to come, Nor height, nor depth, nor any other creature, shall be able to separate us from the love of God, which is in Christ Jesus our Lord."*[46]

This struggle back and forth is traumatic, yet my concern over it led me forward and let me know that hope was not lost.

I do not know your loss, but my grief will never leave me. Every day I remember my children, all of them. Some days I am hit harder than others. It has changed over time. I have a new perspective on life and on love. Before I had children, I thought I understood love. Each time I heard I would be a father again, a new love was born in my heart and it is more than I could have imagined. When I first held my son moments after his birth, it surprised me when my heart swelled with more love. This exact thing happened again when I first held my adopted daughter. As our children age,

---

[46] *Romans 8:38-39 KJV*

love intensifies and matures. What surprised me most about Caleb's death is how much more love I found. My best analogy is all the love that would have gradually grown over his life suddenly matured in an instant.

A similar process occurred in my faith journey. I thought I knew what faith in God looked like until extremely difficult times came. Hardship burns away the excess; takes away the fluff we think is real. From the ember of what remains, a renewed, purified, faith can emerge. For some, like me, hard questions need to be explored. What I have now, though, is a faith beyond what I ever imagined. There is nothing more dangerous to a person's faith than one who knows he is right. God uses the doubt to make Himself known.

Acknowledgements

Donna - Thank you for your love and support, despite the bombardment we have endured. You are an amazing friend, wife, and mother. This book would never have occurred without your help, encouragement, and putting up with my long nights and weekends writing. Thank you for reading multiple terrible drafts and thinking I still had something worth writing.

David, Greg, Jay, and Jeff - I am grateful to have close brothers who supported and cared for me through multiple catastrophes. You have been in the trenches with me a long time. Thank you for always being available and putting up with me when I was not easy to be around. I am blessed to have you and forever indebted to you.

Will - You shaped my spiritual life more than any other. Thank you for listening to me and taking me in when I had nowhere else to go. You have been a fantastic mentor and friend who leads by example. Your family is my family and I love you all.

Oak Leaf Church - In the worst of times, you were present. Thank you for planning everything when I could not, the thousands of hugs, and backing off when it was needed. Thank you for picking up my slack, caring for my family, and constantly looking for ways to help. I will never be able to repay your love or forget all you have done.

Chris and Ben - I cannot express how difficult going back to work would have been without your friendship, understanding, accomodations, and assistance. Without this help, I would not have been able to transition back to work. Thank you for your ongoing patience through my difficult days.

Leave a Review

Thanks for reading *God is in the Doubt*! Your support makes it possible for this independent author to continue creating.

If you liked what you read, please **leave an honest review** wherever you bought this book! Your feedback is invaluable.

About the Author

David, a math and computer nerd since puberty, lives and works from his Atlanta home with his wife of 30 years and their daughter. Whenever he is not home, he is often camping around North Georgia or pondering Einstein's time dilation formula while at the beach. Despite long hours writing software and books, David keeps his faith and family the most important part of his life.

Website:	DavidALloyd.com
Twitter:	@davidlloyd
Facebook:	facebook.com/david.a.lloyd.01
MeWe:	mewe.com/i/DavidALloyd
Gab:	gab.com/DavidALloyd
LinkedIn:	linkedin.com/in/TheDavidALloyd

Made in the USA
Columbia, SC
14 June 2021

40187152R00089